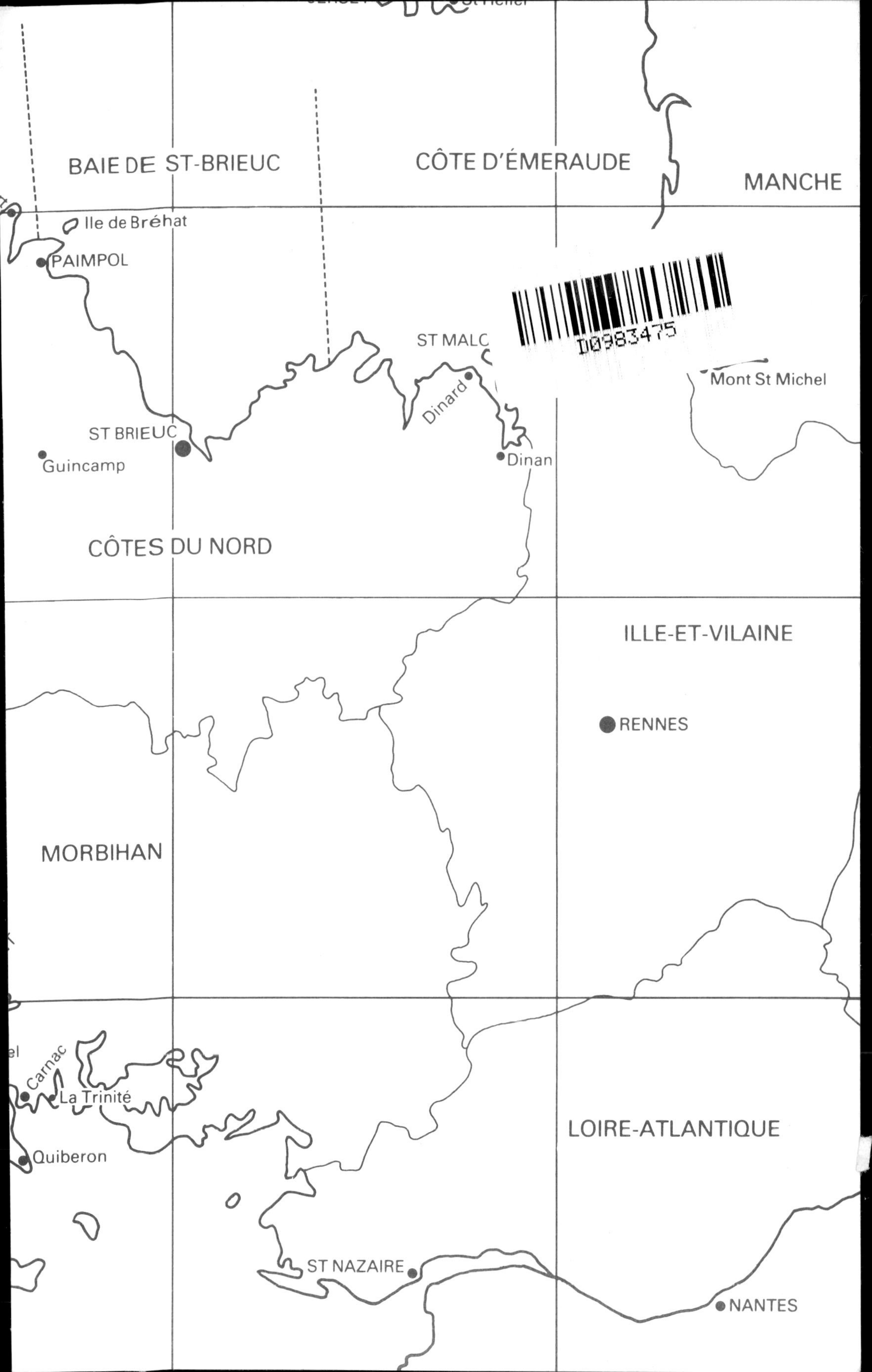

St Helier
BAIE DE ST-BRIEUC
CÔTE D'ÉMERAUDE
MANCHE
Ile de Bréhat
PAIMPOL
ST MALO
Mont St Michel
Dinard
ST BRIEUC
Guincamp
Dinan
CÔTES DU NORD
ILLE-ET-VILAINE
RENNES
MORBIHAN
Carnac
La Trinité
Quiberon
LOIRE-ATLANTIQUE
ST NAZAIRE
NANTES

The Wreck
of the
Amoco Cadiz

Also by David Fairhall

Russia Looks To The Sea

The Wreck of the Amoco Cadiz

DAVID FAIRHALL and PHILIP JORDAN

STEIN AND DAY/*Publishers*/New York

First published in the United States of America in 1980

Printed in the United States of America
Stein and Day/*Publishers*/Scarborough House
Briarcliff Manor, New York 10510

Fairhall, David.
The wreck of the Amoco Cadiz.

Includes index.
1. Oil spills—Atlantic coast (France) 2. Amoco Cadiz (Ship) I. Jordan, Philip, joint author.
II. Title.
GC1321.F34 1980 363.7'394 80-17512
ISBN 0-08128-2743

Contents

List of Illustrations

Acknowledgements

A GREAT MANY PEOPLE gave unstintingly of their time, their professional knowledge, their encouragement and interest and their eyewitness accounts in helping us to put this manuscript together – too many to mention them all by name. We hope they will accept our sincere thanks and indebtedness. In particular we valued the assistance of:

the staff of the Marine Biological Association of the UK in Plymouth, especially Drs Alan Southward and Gerald Boalch; Dr Lucien Laubier, Director of the Brittany Oceanological Centre near Brest; Peter Hope-Jones of the RSPB; Dr Claud Chassé of the Marine Studies Institute of the University of Western Brittany;

Captain Richard Emden, head of the CNIS and Commander Hugh Brunner, the Coastguard's representative on the Marine Pollution Control Unit; Cecil Creber, Principal Marine Surveyor for the South East of England and his colleagues in the Department of Trade, Betty Powell and Liz Drummond;

Dr Ian White and Elsbeth Sutherland of the International Tanker Owners' Pollution Federation; the Chamber of Commerce and Industry in Brest;

Captain Dickson and Margaret Bird of Shell; Mr Ralph Maybourn of BP; Guy Marchand, Maritime Counsellor and Nicole Courter, information officer of the French Embassy in London;

Yves Menez and Daniel Yonnet of *Ouest France* Newspaper; Jack Starr and Stephen Jessel of the BBC office in Paris; Paul Treudhart of the Associated Press in Paris and Mark Webster of *The Financial Times* in London;

the people of Finisterre and the Côtes du Nord and in particular Alphonse Arzel and the citizens of Portsall.

Finally we should like to record our gratitude to the staff of the Liberian Bureau of Maritime Affairs, especially Captain David Bruce in London and Dr Frank Wiswall in New York, because without their generous co-operation the book in its present form would have been impossible.

Prologue

WITH A GALE blasting straight off the Atlantic, the rocky coastline of the Brittany peninsular is an awesome sight. At dusk on the evening of Thursday 16 March 1978, the gale had been blowing for some days. And it was nearly high tide. Enough to prompt Jean Gouzien, deputy lifeboat skipper in the little fishing village of Portsall, to walk out to a spit of high ground from which he could face the wind and scan the waves rolling in through the reefs.

But what he saw hardly made sense: the lights of a large ship – perhaps two ships – only just beyond the Men Goulven rocks, where no sane captain should have been on a night like that. His first instinct was to run to the telephone and call up the local marine radio station. But they seemed to know nothing of any ships in distress. Nor did the search and rescue co-ordination centre further up the coast. So Gouzien went back to the shore, where this time he recognised the familiar silhouette of the Brest-based salvage tug *Pacific*, evidently towing a tanker. All appeared well after all, and the lifeboat skipper called the search and rescue centre back to apologise for raising a false alarm.

In fact, Jean Gouzien was the first eyewitness ashore of what was to become the biggest oil pollution disaster the world had seen.

An hour later, the giant 230,000 ton Liberian crude oil carrier *Amoco Cadiz* grounded on the Roches de Portsall and began to tear herself to pieces on the jagged underwater reefs. Her cargo tanks were instantly ruptured, letting loose a *marée noire* – a black tide – on mile upon mile of coastline, poisoning marine life, killing oysters from some of the best known beds in Europe, fouling seabirds, and leaving the famous white sands ankle deep in evil-smelling crude. It would take an army of men months of painstaking, filthy work even to restore a superficial cleanliness. And a year later the legal battles in the American courts over who should meet the one and a half billion dollar claims for damage had scarcely begun.

This book is built around a detailed account of how the *Amoco Cadiz* ended up on those rocks: bringing together for the first time both sides of the story, the view from the bridge and from the lookouts and radio rooms ashore. Thus far it emerges as one of the strangest and most dramatic sea stories ever told. But our story goes beyond the shipwreck itself to the battle the French waged as the black tide came ashore. And it enlarges this epic disaster into the perspective of our disturbing economic dependence on oil, and its routine transport in clumsy supertankers only a few miles from our shores.

That story starts with the ships and their cargo: by no means the straightforward homogenous product its name implies; and ends among the intricacies of international maritime law. It examines the possible lasting effects of pollution on our coastal environment and asks whether such disasters could be prevented if, for example, ships were controlled from the shore as aircraft are from the ground. And it examines above all how best the French – and any other nation with a vulnerable coastline, like the south coast of England or the eastern seaboard of the United States – can prepare to fight the black tide next time it flows. Because it is beyond doubt that there will be a next time. The only questions are where, and when.

1/Oil and Water

TWENTY-FIVE YEARS AGO, if you had asked a young boy to draw a picture of a ship – just any vessel which to him symbolised ships in general – he would probably have drawn a passenger liner, bristling with tall funnels. Today's child might well choose instead the by now familiar, deeply laden bulk of the supertanker.

Ships have of course been carrying oil around in tanks since the late eighteen hundreds, when the 2,300 ton *Gluckauf* was launched from Tyneside. But the giant supertanker is a creation of the past quarter of a century. In that relatively short time she has come to dominate the maritime scene. Great new shipyards have been laid out to pre-fabricate her deep, square sections; new channels and harbours dug to float them. To prevent her colliding with her clumsy sisters, equally incapable of any sudden change of course or speed, a pattern of maritime highways has been laid down through congested waters like the Dover Strait and the approaches to New York, complete with central reservations to keep the traffic streams apart.

For some people, most notably Aristotle Onassis, she has brought spectacular profits in a glorious demonstration of the economies of scale. For others she has meant nothing but trouble – polluting holiday beaches with her black cargo, demanding that locks be widened and waterways dredged, making old established shipyards commercially obsolete, upsetting the traditional rhythms of seagoing life by the speed with which she loads and discharges at remote automated terminals, rewarding a single navigational error with disaster.

This is the breed to which the *Amoco Cadiz* belonged – at once the simplest and biggest of ships. So big that some crews use bicycles to get from one end to the other. The long hull's design is brutally functional. It has an almost uniform cross section, with no sheer or curve to the deck line. Inside it is boxed off into cargo tanks, with a bulbous-bowed forepeak welded on at the front end and an engine room, propeller and rudder at the stern. Above deck the only striking features are usually the tall double-winged bridge and some sort of streamlined funnel, both of them right aft.

The *Amoco Cadiz* herself had a deadweight tonnage – that is a cargo carrying capacity – of 234,000 tons. On her last voyage from the Persian Gulf she was actually loaded with a mixture of Arabian and Iranian crude oil totalling 223,000 tons. So by the time she set sail from Kharg Island she was deep in the water, approaching her maximum draught of twenty metres. And that meant that much of her long hull (from end to end it measured 334 metres, or more than 1,000 feet, including the bulbous underwater bow) was liable to be awash if she were rolling through a heavy sea. Her propeller, driven by diesel engines mounted in the stern, would then be well below the surface where it was most efficient. In good conditions it gave her a maximum speed of just over fifteen knots.

In short, she was big, clumsy and relatively slow. But not uniquely big by modern standards. In fact no more than a typical supertanker. The world's fleets contain about 700 vessels of 200,000 deadweight tons and over, mostly tankers, of which more than a hundred are in the 300,000 ton class, Ultra Large Crude Carriers (ULCCs) as opposed to mere Very Large Crude Carriers (VLCCs) like the *Amoco Cadiz*.

To explain the evolution of such monsters one need only draw a graph showing the relationship between the cost of transporting a ton of oil from, say, the Persian Gulf to Rotterdam, and the size of the vessel carrying it. As the size increases the cost per ton falls, steeply at first, then flattening out as the tonnage reaches about 200,000.

In the mid-1950's, the typical tanker was still a 16,000 tonner with bridge amidships and machinery aft, an arrangement that had changed little since the pioneering *Gluckauf* was designed seventy years earlier. Vessels like the *Amoco Cadiz* became standard at a later stage because a 200,000 tonner was expected to be the largest size that could be squeezed through the Suez Canal in ballast on a voyage from Western Europe to the Persian Gulf. Loaded, they would return round the Cape of Good Hope.

In time, the big oil producing countries in the Middle East may develop their own refineries, and it will be petroleum products like gasoline, fuel oils and chemical feedstocks that will mainly be transported by sea, in rather smaller tankers, but for the moment, the bulk of the 1,500 million tons of oil that moves across the oceans each year is crude, just as it is pumped up out of the ground. It is the chemical wreckage of ancient life: sulphur, hydrogen, oxygen, nitrogen, organic oils, fats and fatty acids, heavy metals like nickel and vanadium, and more complex carbon compounds synthesised in the tissues of extinct organisms which were buried millions of years ago.

In the ages that have passed, this *mélange* has been compressed, heated,

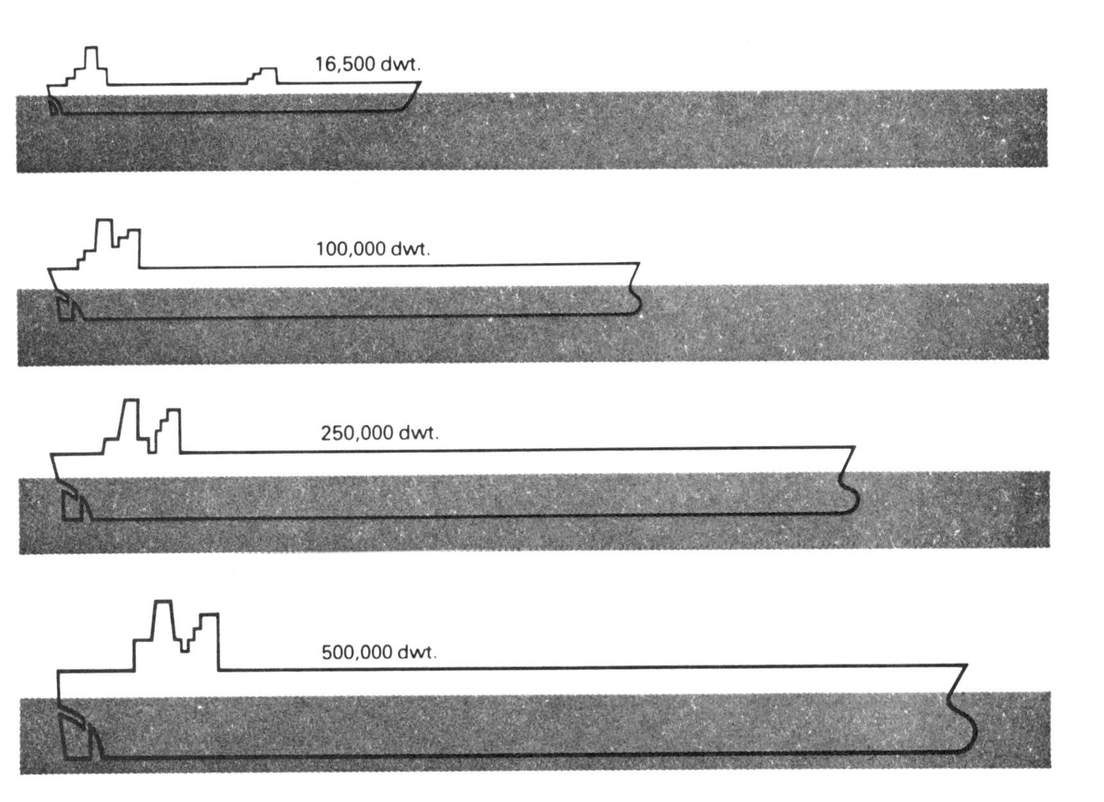

As tankers have grown from the modest 10–20,000 tonners of the 1950's to the half-million tonners of today, their draught has increased until they can scarcely find enough depth of water in the approaches to many of their terminals.

partly oxidized, and jumbled together to produce a complicated substance which has defied precise analysis since the beginnings of organic chemistry. It can perhaps be best described as a complex structure of hydrocarbons, of different molecular weights, in which three main chemical groups – the aromatic, the napthenic and the paraffinic – predominate. The ingredients range from simple, but volatile, components such as benzene, tuolene and xylene – which are known to be highly toxic to marine life – through complex substances, including known carcinogens – the so-called PNAHs – which might hold dangers for man, to intricate waxes and asphalts which cannot yet be properly analysed.

In more than 1,500 known oilfields there have yet to be found two crude oils which are exactly the same. On a conservative estimate there may be more than a million different chemical compositions according to the place and strata from which the oil came. Thus crude oil from Kuwait is as different from North Sea crude as Iranian crude is from that of Venezuela. And one of the things which can mark them apart is the amount of the toxic volatile fractions, often as high as thirty-five to forty per cent.

Oil, of which the world consumes more than sixty million barrels a day, has been oozing into the seabed by natural means almost since it was first formed in the Carboniferous era, some 350 million years ago. No one knows exactly how much oil seeps up through the seabed, but estimates vary between 60,000 and 600,000 tonnes every year: enough at the lower figure to have put 210 trillion tonnes of oil into the water over the millenia. And in addition to this seepage, hydrocarbons are also produced naturally in massive quantities by organisms living in the ocean: some of these are identical to those in oil and some chemically quite different.

And in the past hundred years, man's deliberate exploitation of oil has added massively to the amounts getting into the sea: it comes from the discharges, spills and effluents of refineries; from oil terminals and petrochemical plants; from offshore wells and production platforms; from oil waste; from atmospheric pollution transferred by rainfall; and, importantly for our purpose here, by accident and deliberate act from the 55,000 or more tankers which ply the world's shipping lanes.

If the calculations for natural seepage were correct, that alone would have been sufficient by now to have covered the oceans of the world with oil to a depth of two feet. But it has not happened, for the simple reason that the sea has learned to live with the oil. Aided by its sheer size, 140 million square miles, most of it over 3,000 feet deep (or 328 million cubic miles with four billion tonnes of water in each) the sea is a relatively stable ecosystem with great powers to resist, adapt or recuperate developed over millions of years; it is made up of a grouping of physical, chemical and living subsystems interlocked and in constant flux to maintain the ecological *status quo*.

The central question, however, is whether man's continuing deliberate and accidental contributions to the hydrocarbon levels in the ocean may some day push this ecosystem out of equilibrium with dramatic consequences both for the marine environment, the creatures which live in and around the sea, and for man himself.

Again, no one knows exactly how much oil gets into the sea from all sources but in 1975 the National Academy of Sciences in Washington produced the conservative estimate of 6.11 million tonnes per year.

It thought 0.6 million tonnes (9.8 per cent) came from natural seepage; 0.6 million (9.8 per cent) from atmospheric fallout; 0.08 million (1.3 per cent) from offshore production; 0.8 million (13.1 per cent) from coastal facilities such as sewage plants, refineries, etc; and 1.9 million (31.1 per cent) from rivers and urban area run-off.

As for transportation, it thought tanker operations such as tank washing at sea, loading and discharge spills came to 1.33 million tonnes (at 21.8 per cent, the second biggest man-made source) and other transportation activities at 0.6 million (9.8 per cent); while tanker accidents alone probably contributed only around 0.2 million tonnes a year (3.3 per cent). (A survey by the Exxon oil company suggests that the total from all these causes is matched each year by the natural production of biological hydrocarbons in the ocean).

Figures collected by the International Tanker Owners Pollution Federation from insurers showed 'human error' the biggest cause of a tanker oil spill at forty-six per cent of all cases studied; with equipment failure next at 34 per cent. The most significant equipment failure was a leaking valve (sixty-four per cent) which in some cases was simply because it had not been properly turned off; and the most common human error was 'inattention', forty per cent of all cases.

All these figures are controversial and constantly argued over, but a report from the New York-based Tanker Advisory Centre early in 1979 suggested spills from tankers had increased by 300 per cent between 1974 and 1978. It said spills had gone up from 67,000 tonnes in 1974 to 260,000 tonnes in 1978. Though the figure for 1978 would have been enlarged by the single addition of 200,000 tonnes from the ill-fated *Amoco Cadiz*, the figure for the year before, 1977, was still 204,000 tonnes.

When a spill occurs, a number of physical, chemical and biological processes going on naturally in the ocean act on the oil. Some of them are most important immediately after the spill; others become increasingly important as time goes on.

As crude oil hits the water it begins to spread very rapidly. The water works with surface active agents in the oil, such as the sulphur and oxygen compounds, and, after the first hour or two, slick size for most spills is generally controlled by the relationship between surface tension and its viscosity: these compounds often spread so quickly that a chemical 'flash' can shoot across the water, making a slick look much larger than it is. In practice, the slick is most likely to be made up of huge areas of thin oil film, which might be a layer as thin as a single molecule, or iridescent sheen, while the bulk of the oil is held together in smaller floating islands of much greater thickness.

As the oil spreads, a dominant process which will wane gradually over a week to ten days, the most volatile lighter fractions begin to evaporate off. In a large slick this can be a major cause of oil disappearing and the amount evaporating can be as high as fifty per cent. It can go on for some time but in most cases it takes place within the first few hours. Calculations on the *Torrey Canyon* spill showed that twenty-five per cent of its oil evaporated in less than twenty-four hours.

But at the same time these light fractions often contain the more soluble elements which, still in a highly toxic form for marine life, pass straight into the water. The two processes are competitive, the same fractions being removed by both.

What is left then spreads less rapidly because the disappearance of the lighter fractions has made it more viscous, and it is open to attack by both photochemical and bacterial means. The floating oil is degraded by atmospheric oxygen and the effects of solar radiation, as well as by more than ninety known organisms which are capable of living on hydrocarbons. It has been calculated that a shoal of one organism at 2,000 to the square centimetre and ten metres deep over an area of a square kilometre could take up three tonnes of oil a day.

But other factors come into play. If the sea is rough at the time of the spill, the churning action of the water will start to disperse the oil into fine droplets, often spread to a great depth and over large areas.

Such droplets are easy targets for photochemical and bacterial attack but if the oil quantity is large it does not take much sea action to turn it instead into a water-in-oil emulsion. Great balloons or globs of this material, which marine biologists have dubbed 'chocolate mousse' and which can contain up to seventy or eighty per cent water, will distribute temselves, floating almost submerged and gradually congregating into great mousse sheets.

Sometimes the mousse is only a few inches thick, but spills have been reported in which it was so stable it was possible to stick an oar up vertically in it without it falling over. Bacterial attack is slowed down due to the slow diffusion of oxygen and nutrients to the inner layers, and only the outer layers are available for photochemical degradation, so these emulsions can persist for some time, forming long windrows and rafts, interspersed with sheen, which can be carried along by wind or tide and washed ashore, to engulf rocks and weeds and mix with sediment and beach debris.

Where the oil is untouched by man, degradation will then take some considerable time. Even under favourable conditions, according to some scientists, there may be persistent spilled oil whose half-life must be measured in terms of years. All that is volatile, oxidizable or readily biodegradable disappears, but the residues, which in a slick away from the shore may often be found floating in the sea as oil sludge or tar balls up to ten centimetres in diameter, may still represent between ten and thirty per cent of the original oil. These lumps, which may be either soft or brittle, sometimes become covered with sediment or marine growth and sink to the bottom, which is believed to be the fate of a great deal of oil in the ocean.

2 / Signposts to Disaster

FOR MOST PEOPLE, it was the loss of the *Torrey Canyon* in 1967 which first demonstrated the devastation that a big oil spill can bring: beaches swilling with black sludge; tens of thousands of sea birds slowly dying, of which a mere one hundred were successfully cleaned and returned to the sea; fishermen forced to choose between oil pollution and a detergent remedy which, in those days, probably did more damage than the problem it was supposed to cure.

The *Torrey Canyon* was a 121,000 tonner, flying the Liberian 'flag of convenience' on behalf of the Barracuda Tanker Corporation in Bermuda and the Union Oil Company of California, to which she was chartered. Her master was an Italian, Captain Pastrengo Rugiati.

For those who followed the subsequent Liberian inquiry, the most shocking aspect of the disaster was the almost trivial nature of the navigational errors that caused it. The crew was not struggling with any hazard such as a North Atlantic gale. The big tanker scarcely altered her monotonous course from the time she left the Canary Islands on the west coast of Africa until 0850 on 18 March when she struck the Seven Stones reef off Britain's Land's End, her engines full ahead, her master fumbling with the selector switch of the automatic helmsman. Nor was it a case of the navigator having lost his way in bad visibility. The Seven Stones light vessel was clearly visible only two and a half miles from the Pollard Rock on which she actually grounded. Indeed, according to the Liberian board of investigation's report, the light vessel was used, together with a radar range, to fix the tanker's position only ten minutes before the disaster:

> The master, realising that he was closer to the Seven Stones than expected (only about 2.5 miles) decided to come left to 000 degrees. He testified that he was prevented from any further alteration of course to the left by the presence of one of the fishing boats (identity unknown) on his port side. He also testified that there was no vessel on his starboard side which would have prevented a right turn at that point.
>
> About 0848 another position was obtained by the third officer, using radar range and visual bearing to the light vessel. The distance was only about 2.78 miles.

> The master knew that his vessel was perilously close to the Seven Stones and he shouted to the helmsman to come hard left. The helmsman immediately went to the steering stand and put the wheel over hard left, but saw there was no response. The vessel remained on course 000 degrees.
> He promptly called out to the master, who ran to the wheel and saw that the selector switch was in the control position, not 'hand'. The master immediately turned the lever to 'hand' and the bow of the Torrey Canyon began to swing left. According to the course recorder graph she reached a heading of 350 degrees when at 0850 the vessel struck Pollard Rock and came to a sudden stop, hard and fast aground. . . . There had been no orders to the main engines prior to the stranding, and the *Torrey Canyon* hit at full speed of approximately 15.75 knots.

The Liberian report criticised Captain Rugiati for three mistakes: keeping his ship on automatic steering while near the Isles of Scilly, failing to reduce power before the stranding, and failing to alter course to starboard, to pass eastward of the reef in good time, when a safe alteration to port was prevented by the nearby fishing vessel. Cancellation of his master's ticket was recommended.

As *The Guardian* newspaper put it in an editorial comment: 'One Italian sailor made three mistakes, and so a thousand miles of French and English coast are threatened by the *Torrey Canyon's* oil.' Yet however just the board's censure, Captain Rugiati had become a scapegoat for the failure by many others – seafarers, shipowners, governments – to meet the threat posed by these giant ships with a commensurate effort to improve their safety, whatever it might cost. Nothing illustrated this more cogently than the Italian skipper's explanation of why he took the short cut inside the Scilly isles in the first place. It was partly, he said, because his owners' agents had urged him to catch the evening tide at Milford Haven that day. A safe course outside the islands would have taken twenty-nine minutes longer, the board calculated – and he would have caught his tide.

To be fair to both governments and the international oil industry – both of them inclined to be neurotically defensive on these issues – a great deal of sound work on the prevention of tanker accidents and oil pollution generally had been done before March 1967. For example the International Convention for the Prevention of Pollution of the Sea by Oil dates from 1954, controlling deliberate rather than accidental discharges of oil. The big tanker casualties obviously do a great deal of damage because the oil spill is concentrated over a few hours or days, but a much larger quantity of oil actually reaches the oceans from ships washing out their cargo tanks or bilges. Tank cleaning is a regular – and potentially dangerous – part of every crude carrier's seagoing routine as she returns to the oilfields in ballast. The 1954 convention therefore sought to prevent their doing this within fifty miles of land and make sure that the oily waste was heavily diluted with water before it was pumped overboard. It is under these rules, which have since been tightened and extended, that ships are sometimes prosecuted for causing oil pollution – when for example a maritime patrol aircraft spots a slick trailing from a vessel in coastal waters.

Another absolutely fundamental measure, the establishment of one-way traffic separation schemes to reduce the number of collisions in congested areas, was the subject of discussion and experiment in the early 1960's, with Shell International Marine leading the campaign for enforcement. But in those days the debate tended to be rather abstruse and technical. After the *Torrey Canyon*, politicians and public quickly got the simple message – the industry would say oversimplified – that big tankers were dangerous, and began to press governments and oil companies for more safety measures. The professionals, meanwhile, had learned a lot of hard practical lessons about cleaning up and dispersing oil – or rather about their inability to deal with it on such a scale.

One obvious essential was to provide adequate financial compensation for major oil pollution damage like this and for the cost of clearing it up. Before 1967 and the Torrey Canyon there was no appropriate legal regime for this. The British Government put in a claim for £2.9 millions, and the French Government demanded a similar amount. But lawyers for the Barracuda Tanker Corporation ignored them until the *Torrey Canyon*'s sister ship, the *Lake Palourde*, on her way from the US West Coast to the Persian Gulf, misguidedly put into Singapore – reportedly for nothing more vital than a couple of coils of wire. The Singapore High Court bailiff, under instructions from the Treasury Solicitor's Office in London, slipped aboard and nailed – well, pasted – a write on one of her masts 'in the exercise of the Admiralty jurisdiction of the High Court of Singapore'. With their ship under arrest, the owners were prepared to deal, but even then the two governments only recovered about half their claims.

There was an almost romantic touch of legal ritual about the affair, and the Treasury Solicitor's Office certainly deserved credit for initiative, but it was hardly a satisfactory precedent for future settlements. Tanker owners and maritime governments therefore set their lawyers to devise a regime that would guarantee insurance against pollution costs, clarify the industry's liability, and then set what was hoped would be a generous limit upon it.

The voluntary industrial schemes came first: TOVALOP (Tanker Owners' Voluntary Agreement Concerning Liability for Oil Pollution) in 1969 and CRISTAL (Contract Regarding an Interim Supplement to Tanker Liability for Oil Pollution) in 1971, with the international oil companies appearing in both because they own something like forty per cent of the world's tanker tonnage themselves. TOVALOP provided owners with mutual insurance cover of up to 16.7 million US dollars for any one incident, and the oil companies would then top this up to a total of $30 millions (since raised to $36 millions) through CRISTAL.

As for the governments, some, like the United States, preferred to rely on national law. Others, including the British and French, chose to work through the ponderous machinery of IMCO (Intergovernmental Maritime Consultative Organisation) to back TOVALOP with the Civil Liability Convention and CRISTAL with the Fund Convention, but only the first of these governmental agreements was in force by March 1978 – even for its signatories, which included France and Britain but not the USA.

IMCO is the only major United Nations organisation located in London. More than one hundred nations, including all the major maritime countries, are now members. As one would expect, it suffers from a good deal of diplomatic inertia, but for all that its work is invaluable. Without it, international seafaring would be a chaotic and much more dangerous occupation.

It is IMCO, for example, that nowadays lays down such fundamental rules as carrying a red sidelight to port and a green one to starboard, or sounding two short blasts on a ship's siren before turning to port, or giving way to a vessel crossing from the right – although some of these are of course derived from law or practice that greatly pre-dates the UN organisation. Even the old rule about steam giving way to sail still appears in the IMCO collision regulations, albeit much modified by modern reality and the average yachtsman's sense of self-preservation.

Successive IMCO conventions on the Safety of Life at Sea have also laid down minimum standards of ship construction, engineering, navigational and safety equipment, all of which then become subject to certification and inspection – in Britain by the Department of Trade and in the US by the Coastguard. IMCO conventions are not in themselves maritime law, but they bind countries which accede to them to enforce the rules through national legislation once a certain number of signatories, controlling a specified proportion of the world's shipping tonnage, have ratified them. This can take many years, but the process has recently been speeded up by setting target dates for implementing various measures.

Most importantly, the governments which do sign are not just representing the traditional maritime nations. They include those like Liberia and

Panama, which operate 'flags of convenience' for a large proportion of the world's tanker fleets.

The Organisation's immediate response to the *Torrey Canyon* disaster was to call its council together in extraordinary session and lay down an impressive programme of investigation, to establish what could be done to lessen the hazards of oil transport by sea, under broad headings of prevention and cure. Looking back it is still an impressive programme, because virtually all the international measures that have actually been taken since then, plus some that are still vigorously debated, were outlined after that conference in May 1967.

The council agreed to study as a matter of urgency:

1. whether sea traffic lanes should be established, and if so whether they should be voluntary or compulsory;
2. whether large ships carrying oil and other potentially dangerous cargoes should be prohibited from certain sea areas and certain routes;
3. whether such vessels should be made to carry additional navigational aids;
4. whether shipping in coastal waters should be guided by shore radio, and if so how it should report its presence;
5. whether international standards should be established for the training and qualification of crews, particularly on large tankers;
6. operational rules for automatic pilots;
7. improved design and construction standards to limit oil spills in the event of stranding or collision;
8. better ways of stopping big heavy ships;
9. the need to improve look-out systems;
10. international contingency plans for cleaning up oil spills;
11. more stringent controls on the deliberate discharge of oil;
12. chemical dispersants that did less damage to the living marine environment;
13. the extent to which governments should intervene between ship-owners, insurers and salvage companies to protect their coastlines;
14. compulsory insurance against pollution damage and the recovery of clean-up costs.

Some of these ideas, such as banning supertankers from certain sea areas, made little headway. Some were quickly adopted. For example the first traffic separation scheme was established in the Dover Strait between England and France in June 1967, albeit only on a voluntary basis at first. Others required years of diplomatic groundwork, so that the first IMCO convention on training and certification, aimed at setting minimum standards in countries that did not have their own infrastructure of maritime law and tradition, was not ready for signature until 1978.

IMCO was urged on by the major oil companies, who took pride in the generally high professional standard of their own tanker operations and resented being tarred with the brush of indiscriminate public criticism. The oilmen anticipated the liability and compensation conventions with their voluntary insurance schemes. They developed less toxic chemicals for dispersing oil, and mechanical devices for skimming it off the surface of the sea. They joined forces with the British Government's scientists to see whether there was a more efficient way of setting fire to a standard tanker than dropping bombs on it – as the Royal Navy did with the *Torrey Canyon*. But for all this energetic work, the impetus imparted in 1967 would probably have been allowed to fade had the public's short memory not been jogged every few years by a succession of similar disasters, each one pointing to some new moral.

In 1970, two tankers carrying, between them, far more even than the *Torrey Canyon*, collided only a few miles off the holiday beaches of the Isle of Wight, on the south coast of England. By good fortune, only 6,300 tonnes of oil was spilled and once the local rescue services were alerted, they proved to be efficiently organised. But there was a great deal of public alarm nevertheless.

It was noted that these vessels, too, were flying the Liberian flag of convenience; one was manned by Greeks and the other by Chinese. And as the full story of the casualty emerged it became clear that the two tankers had been alongside each other, on slowly converging courses, for half an hour before the collision. One vessel, the *Pacific Glory*, then lay motionless for two hours, without summoning assistance, until she exploded and burnt with the loss of fourteen lives.

Early in the following year, fifty-one seamen died further up Channel in the marine equivalent of a motorway pile-up. First the Peruvian freighter *Paracas* collided with the tanker *Texaco Caribbean*. Then two more ships, the German freighter *Brandenburg* and the Greek coaster *Niki*, struck the sunken wreckage and went down themselves.

In 1975, another collision on the other side of the Dover Strait between HMS *Achilles* and the tanker *Olympic Alliance* came as a reminder that naval vessels are not immune from casualty, for all their navigational discipline. This time 2,100 tonnes of oil swilled past the white cliffs of Kent.

The *Amoco Cadiz* apart, 1978 was a bad year. On 6 May, with events in Brittany still much in the news, a French bulk ore carrier, the *Roseline*, sliced right through the Greek tanker *Eleni* V only six miles off the Norfolk coast in eastern England. Much of the 5,500 tonnes of oil that was spilt came ashore within forty-eight hours. But another large quantity was left trapped in the severed tanker's upturned bow section, drifting with the tide in sight of the cliffs.

All the local communities (many of them holiday resorts preparing for the new season) wanted was for somebody – anybody – to come and take it away. But this casualty was to teach some new lessons. The *Eleni* V, it turned out, had been loaded with heavy fuel oil, a foul, treacly substance that was not susceptible to the chemical dispersants that had been used successfully to treat crude oil slicks at sea. To make matters worse, the heavy oil would need heating before it was liquid enough to be pumped out of the capsised bow.

After a lot of effort, tugs and divers managed to get the wreck under control, but there the salvage operation stuck. Attempting a long tow round the north of Scotland was considered too risky. The hulk might have broken adrift, or sunk completely, and then fractured one of the oil and gas pipelines on the seabed. Another scheme was to dredge a sort of harbour for it on the beach and contain the oil with a boom, but this was judged impractical.

In the end they just towed the hulk further offshore and the Royal Navy had a fine day out blowing it into as many pieces as possible to release the remaining oil. Seven months later, in an official post mortem on the incident, the Government admitted there was not much else it could have done:

> The plain fact is that as yet no one – whether in the United Kingdom or overseas – has demonstrated a technique for dealing satisfactorily at sea with spills of heavy oils. Should it be necessary to deal with another spill of heavy oil in the future, the appropriate approach may well be to undertake a restricted spraying operation using approved dispersants and if, on scientific appraisal, this confirms the relative ineffectiveness of dispersants in the particular circumstances, to concentrate on cleaning up any oil that comes ashore by mechanical means, which is the main recommended method of dealing with all types of oil on beaches.
> (*Eleni* V – the Government's reply to the fourth report from the Select Committee on Science and Technology, Command 7429.)

About ten per cent of the oil that is shipped round the British coast comes into the heavy, viscous category covered by this commendably honest but disturbing admission. And British coasts are not the only ones along which such cargoes pass.

The story of the Greek tanker *Christos Bitas,* which struck submerged rocks off the coast of South Wales in October 1978, ended more happily. Not that coastal pollution was avoided. About 2,400 tons of crude oil escaped from tears in her bottom plating, and in spite of a well-organised spraying effort, some of this found its way on to the unspoilt coastline of Wales and Devon, killing at least 2,000 birds in the process. Another 1,200 tons of cargo and bunker fuel was sunk deep in the Atlantic when the ship was eventually scuttled. But the salvage operation which initially kept her afloat and enabled nearly 31,000 tons of crude plus most of the bunkers to be offloaded into other tankers, was an object lesson in how much progress had been made in the eleven years since *Torrey Canyon.*

To begin with, there were disturbing echoes of the earlier disaster, though on a smaller scale. The *Christos Bitas* had come round from Rotterdam, part loaded with 35,000 tons of Iranian crude, and was heading up the Irish Sea when she too cut a navigational corner too fine. Four and a half miles inside the Smalls lighthouse, instead of outside as one would expect, she ripped her bottom open on the Hats and Barrels reef. First news of the casualty came from the Ilfracombe coastal radio station in a message to the local Coastguard, the Royal Navy's Flag Officer in Plymouth, and Lloyd's of London:

> Following received from Greek tanker *Christos Bitas*/SXAB at 16.34 on 500 kHz. Greek flag tanker grounding 4.5 miles east of Smalls. Damaged bottom leaking oil. At present only require assistance to avoid oil pollution.

Fortunately, the Greek vessel immediately freed herself from the rocks (which would probably have prevented other tankers coming alongside to lighten her) and the Coastguard requested her to steam slowly seawards. The South Wales anti-pollution plan was activated, an operations room established at St Ann's Head alongside the Coastguard rescue centre, and the Department of Trade's marine emergency information room manned in London. But to begin with, the unknown captain on the tanker's bridge seemed remarkably unconcerned.

At 18.15 HMS *Hecate,* a hydrographic survey vessel that happened to be in the area, reported that the *Christos Bitas* was turning to port, spilling oil as she went, and had developed a list to starboard. Yet fourteen minutes later the tanker told *Hecate* the leakage had stopped and the ship was safe. At 20.45, the dispersant spraying tug *Exegarth,* which had come out from Milford Haven to help, found herself trailing five miles astern of the casualty as she steamed north. Shortly after 21.00, HMS *Hecate* confirmed that the Greek tanker had simply resumed her voyage to Belfast, in Northern Ireland.

At this point both the Department of Trade and the BP Tanker Company, which had chartered the *Christos Bitas*, radioed the Greek vessel to stop so as at least to reduce the area of pollution that must be covered. Shortly afterwards, at 22.41, the tanker did stop, about forty miles north of the Smalls, and within a few hours it became clear that she was in a much more dangerous condition than her master seemed originally to think. Her starboard list had reached seven or eight degrees and was still increasing. At 04.00 on 13 October she put out a distress call and by 06.30 that morning the Rosslare lifeboat from Ireland had taken off nineteen of her crew.

A couple of hours later, when BP's marine superintendent landed on board by helicopter, having driven across Britain overnight, he found the starboard list increased to fourteen degrees. But if she could be prevented from capsizing long enough, it was still worth trying to offload her cargo at sea. None of the ports round the Irish Sea would give refuge to the crippled tanker anyway. Milford Haven's Port Authority had made its own position clear almost as soon as the casualty was reported.

Beaching her was considered only briefly. Setting fire to her was unlikely to do much good, even if the insurers would agree to it, because the decks were already awash forward. Simply letting her sink in the middle of the Irish Sea would produce an outcry from environmental interests. So it was agreed that two tankers specially equipped – for example with giant inflatable fenders – for transferring oil at sea should be ordered to the scene. High capacity pumps would be airlifted from Rotterdam and Bergen with their crews.

Meanwhile United Towing's powerful salvage tug *Guardsman* had turned up and offered to work on a 'no cure, no pay' basis. She gently towed the *Christos Bitas* towards a point in the St George's Channel where it was reckoned pollution would do least damage and a growing fleet of spraying vessels scurried round trying to disperse whatever oil they could find. At that stage the slick was estimated to measure about ten miles by six miles.

Thick overnight fog made things more difficult, especially for the helicopters. First attempts to correct the damaged tanker's list by pumping her cargo from one side to the other failed because sea water flowed into the ruptured tanks as fast as the oil flowed out. Compressed air simply hissed away until special airtight tank fittings, some of which were made on the spot, aboard the BP tanker *British Dragoon*, were supplied on the morning of 15 October. Nevertheless the first 1,900 tons of crude had been transferred to the *Esso York* during that third night. *British Dragoon* took over from her the following evening and progressive pressurisation of the tanks began to lift the *Christos Bitas* slowly clear of the water. The pumping rate rose to 300 tons an hour. By the early hours of 17 October the *British Dragoon* had more than 12,000 tons aboard and the casualty was once more upright. She needed to be, because a north westerly gale was on its way.

The two ships separated to ride out the bad weather, which gave the salvage teams a chance to catch up on some of their sleep. The same gale drove some of the smaller spraying vessels working off the Welsh coast into shelter. But when it died down only three miles of the hundred or so miles of coastline most immediately threatened seemed to have been polluted. More oil appeared later, but a great deal had evidently been achieved by a spraying and collecting fleet that at one time numbered thirty-five vessels, assisted by aircraft. A total of 27,000 gallons of concentrated dispersant was used, plus 40,000 gallons of the standard type.

British Dragoon was back alongside the Greek tanker by noon on 18 October. Paradoxically, the work of pumping out became more difficult as the damaged ship gradually recovered her buoyancy. Pumps had to be manhandled in and out of the deep tanks. Sometimes men wearing breathing apparatus had to go down with them into the fume-laden darkness. The risk of an explosion was always there.

The success of this part of the operation was finally registered at 02.00 on 22 October. The *British Dragoon* had more than 30,000 tons of oil aboard. The *Christos Bitas* was left with the dregs of her bunker fuel and perhaps 1,000 tons of crude swilling around in the bottom of her undamaged tanks or clinging to the steelwork.

Now the debate about what to do with her started all over again, with the British Government firmly in the chair since its formal intervention under Section 12 of the Prevention of Oil Pollution Act of 1971 (another post-*Torrey Canyon* measure). There was talk of dry-docking her in Falmouth, but that might have been a dangerous business and the local fishermen were none too keen. The authorities on the Clyde said they might let her in there, but only if repairs were carried out in their shipyards. The French were approached about the possibility of towing her through the channel to Rotterdam. Finally the argument was settled by the owners' calculation that even the cost of cleaning their vessel and freeing her from gas would be more than she was worth. It was decided to tow her 600 miles out into the North Atlantic and sink her beyond the edge of the continental shelf.

The job was done by United Towing's 10,000 horsepower salvage tug *Guardsman*, with the 9,000 horsepower *Yorkshireman* standing by. An eight-man salvage party commanded by the company's salvage master stayed on board the *Christos Bitas*. Her boilers were shut down, leaving only an emergency diesel generator to provide some light and enough power to work the steering gear.

The tow started in fine weather on the evening of 26 October. With only a slight sea running, and not much swell, *Guardsman* was able to maintain a speed of about five knots, heading south west for the first couple of days to keep at least eighty miles clear of the Irish coast. British and Irish naval vessels were in attendance.

On the morning of 29 October the convoy altered course to slightly north of west, so as to steer for really deep water from which any oil seepage would be carried clear of Ireland and Scotland by the North Atlantic Drift. But a southerly swell was building up now and *Guardsman* had to ease down to less than four knots to prevent the towline snatching. Even so the line parted on the tanker's foc'sle and it took nearly six hours for the tug to recover her gear and reconnect, this time using the port anchor cable of the *Christos Bitas* at that end.

The weather continued to worsen. By the morning of 31 October the big salvage tug was pitching into a heavy westerly swell. Speed was down to two knots and the tanker astern, like a patient beginning to relapse from a temporary recovery, had once more developed a five degree starboard list. The morning weather forecast was promising anything from gale Force 9 to storm Force 11. In those conditions no progress could have been made even if the towline had remained intact. It was decided to let her go there.

The *Yorkshireman* connected a line to the *Christos Bitas* to hold her bow up into the weather while *Guardsman* recovered her gear and rounded alongside to take off four of the salvage party. The remaining four began a carefully planned scuttling operation. First valves were opened to flood the engine room; then the cargo tank valves to begin flooding the main body of the hull, and the ullage ports to release the trapped air. The final stage is best described in the cool, formal language of the official report:

> At 14.15, with the *Christos Bitas* visibly settling by the stern and starting to list further to starboard, the remainder of the salvage party were taken off by the *Guardsman*. The *Christos Bitas* slowly settled by the stern and rolled over to starboard until she lay with her side almost horizontal, from which position the stern sank, pulling the bow down with it. The vessel disappeared below the sea surface at 15.40 in position 51 degrees 22 minutes N, 18 degrees 13 minutes W. The oil slick caused by her sinking extended about 600 feet by 400 feet at its maximum, and whilst there were a few patches of black oil it very quickly

thinned out and the observers on board the tug were well satisfied that spillage of oil had been minimised.

Ignoring that final touch of propaganda, it must have been a chilling sight even if one was absolutely sure there was nobody left on board.

The beginning of 1979 nearly saw a repetition of the *Amoco Cadiz* disaster, and in direct human terms, what actually happened was far worse. Most of the crew of the 219,000 Greek tanker *Andros Patria*, plus the master's wife and two-year-old son, were lost after abandoning their ship in heavy weather thirty miles off the north west corner of Spain.

Like the *Amoco Cadiz*, the Greek VLCC was bound from the Persian Gulf to Rotterdam with a cargo of crude oil when at 8 pm on New Year's Eve one of her tanks was torn open on the port side. There was an explosion followed briefly by a fire. The shock of the fire must have been heightened for her crew by the fact that they were busy at the time preparing to celebrate the New Year. The master understandably decided to leave while his ship was still in one piece, but as often seems to happen, it was the three who stayed aboard, including the chief engineer and the bo'sun, who survived.

The giant vessel was leaking tens of thousands of tons of oil from the tear in her side. The Galician fishermen began to fear another wave of pollution, only two and a half years after the *Urquiola* exploded in the entrance to La Corrunna harbour. In fact two Dutch tugs, Wijsmuller's *Typhoon* and Smit's *Poolzee*, were able to get lines aboard and tow the casualty away from the coast. From there, they began searching for a refuge which none of the authorities in Spain, Portugal, France or Britain were prepared to offer until they were sure she was no longer a pollution hazard. In this respect the situation was similar to that of the *Christos Bitas*.

For three weeks, a strange company of ships wandered 1,500 miles south and west into the Atlantic, to find calm enough weather to start transferring oil into BP's *British Dragoon*, the 50,000 ton lightening tanker that had handled the operation in the Irish Sea. She then passed it on to the VLCC *British Promise*.

Another fifty people had meanwhile died on 8 January, when the French 120,000 tonner *Betelgeuse* exploded alongside the discharging jetty at Bantry Bay, in southern Ireland. Just before she went up, a frenzied voice, believed to have been that of the harbour pilot, was heard on the radio calling to the others on board to 'get off, get off, for God's sake get off'. It turned out that the French tanker did not have an inert gas system with which to flood her tanks as they were emptied. This particular safety measure was not mandatory when she was built in 1968, the owners explained.

Since the tanker business is a truly international affair, American oil companies and shipowners obviously took careful note of all these casualties in European waters, even when they were not directly involved. But some that were rated locally as disasters reached the wider public in the United States only as statistics confirming a disturbing trend. The loss of the *Torrey Canyon*, the first big event of its kind, obviously made an impact even at that distance. So did the *Amoco Cadiz*, which set all kinds of grim new records and happened to be American-owned even if she was not flying the American flag. But the rising tide of public concern during the intervening years, with its pressure for tougher controls and the steady erosion of command at sea, was encouraged in the United States by a different pattern of events.

In terms of the size, if not the number of oil spills, the American coasts have suffered a lot less than European ones. But to make a useful comparison, one has to remember a number of differences.

To begin with, US imports of oil and petroleum products by sea amount to only about two-thirds the European total. In 1977, for example, US ports handled about 370 million tonnes of imported crude (that is 250 million less than their European counterparts) and seventy million tonnes of petrol, kerosene, heavy oils and other products (that is half the equivalent European figure).

The size of ship which can use American ports is also limited by the generally shallow approach channels. Few, if any, can handle a tanker drawing more than about fifty feet of water, which is roughly the draught of a fully laden 100,000 tonner. Vessels of 150,000 or 200,000 tons are nevertheless used in this trade, but they have to be lightened offshore before moving in to the terminal.

Whereas the smaller total quantity of oil moved by sea directly reduces the pollution risk, the effect of using smaller ships and transferring oil at sea could work in the opposite direction. The oil companies certainly argue, as part of their campaign to persuade us to love the supertanker, that giant vessels are a safer way of transporting crude. Ports which stand deep in bays or estuaries, as some of the American ones do, also pose extra problems.

Two further generalisations are probably worth making about the United States scene by comparison with Europe. One is that even without a *Torrey Canyon* on its own shores, the American public has shown itself peculiarly sensitive to environmental issues like oil pollution and gets a good deal of encouragement from US law and the US courts in the process. The second is that under pressure from powerful Congressional lobbies, the American Administration tends to prefer unilateral legislation – for which the timetable can be set by domestic political considerations – rather than the tedious process of international negotiation and compromise. An example is the 'Superfund' law passed in 1978 to provide for oil spill compensation and govern liability just as the two IMCO conventions do for their European signatories.

In fact seen through the eyes of the British maritime establishment, many recent campaigns to improve safety at sea have resolved themselves into a struggle to stop an impatient US Government going unilateral until some of the smaller nations whose ships actually cause the trouble could be persuaded to take joint action through IMCO. Yet in spite of this, and partly for the reasons sketched in above, it seemed to be a long time before the message of the *Torrey Canyon* really reached Washington.

The Santa Barbara incident rang a few alarm bells in 1969. It involved probably the biggest coastal clean-up operation the United States had so far seen, only a few miles from Los Angeles and therefore well publicised. But it had nothing to do with tankers. The trouble was caused by the blowout of an offshore well, which leaked about 12,000 tonnes of oil into the Santa Barbara channel over the next three months, killing sea birds and polluting beaches in familiar fashion. In one respect, the fact that the oil seeped from the seabed over a long period, it was worse than a once-for-all spill.

But it was the winter of 1976–77 that really stirred things up in Washington, and prompted a major new US initiative to raise tanker safety standards. It was a bad period for casualties worldwide, and no less than eight of them occurred in American waters between December and March of those years. Some of the tonnages were admittedly quite small, yet the cumulative effect was devastating, particularly on the public image of the much criticised flags of convenience:

15 December – *Argo Merchant*, 29,000 tons deadweight, Liberian flag, runs aground off Nantucket Island, spilling 28,000 tons of fuel oil.

17 December – *Sansinena II*, 72,000 tons, American flag, explodes and burns off Long Beach, spilling seventy tons of crude oil.

24 December – *Oswego Peace*, 50,000 tons, Liberian, cracks hull in Thames River, Connecticut, spilling seven tons of oil.

27 December – *Olympic Games*, 61,000 tons, Liberian, grounds in Delaware River, spilling 400 tons of crude.

29 December – *Richard C. Sauer*, 65,000 tons, Liberian, strikes submerged object off Staten Island and spills 800 tons of crude

30 December – Last report from *Grand Zenith*, 32,000 tons,

Panamanian, 'experiencing heavy weather' thirty miles off Cape Sable, Nova Scotia, loaded with fuel oil from Teesport, England, for Somerset, Massachusetts. Ship and thirty-eight-man crew posted missing 9 January; two lifejackets and oil slick found two weeks later 300 miles ESE of Cape Cod.

4 February – *Ethel H.*, 8,000 tons, American, grounds in Hudson River, spilling 1,200 tons of fuel oil.

24 February – *Hawaiian Patriot*, 99,000 tons, Liberian, cracks hull in storm off Honolulu and burns, losing 16,000 tons of crude.

Apart from the special case of the *Grand Zenith*, the first ship in this catalogue of accidents, the *Argo Merchant*, spilt more oil than all the others put together. In other respects, too, hers was the most disturbing story, even though there was no loss of life. The way she blundered on to the Nantucket shoals soon became a familiar cautionary tale on that side of the Atlantic, just as the stranding of the *Torrey Canyon* had on the European side; a standard against which other tales of oil and misfortune could be measured.

The twenty-three-year-old Liberian tanker left Puerto La Cruz in Venezuela on Sunday, 5 December, bound for Salem Massachusetts with a cargo of heavy fuel oil – the thick kind that requires heating coils in the tanks to make it pumpable. Her voyage took her through the Mona Passage, past Cape Hatteras and then north east to Nantucket where she would turn the corner – marked by the Nantucket light vessel – and head in towards Salem.

According to the US Coastguard's Commander Calicchio, who boarded her later, she definitely looked her age, with rust staining decks and ladders. Nor was this just a superficial impression, because she had a history of engine room breakdowns. Her navigational equipment was adequate when it worked. The charts were six years old.

The *Argo Merchant* was the sole vessel owned by the impressive sounding Thebes Shipping Corporation, part of an obscure network of Graeco-American-Liberian companies directed from Brooklyn, New York. Her crew were a mixed bunch, to say the least. The Master, Captain Georgios Papadopoulos, was Greek, like his officers. But the rest of the crew were of various nationalities and some of them could speak neither Greek nor English. On this trip they were short-handed anyway, because a couple of able seamen had paid off in Puerto La Cruz and the Master got no response to his request for replacements. This led to an argument that first night at sea. When Third Officer Nisiotis found he was expected to man the bridge with a recently promoted able seaman who scarcely knew how to steer and an engine room 'wiper' transferred to the deck to make up the numbers, he refused to take his watch with them.

There was more trouble, of a mechanical kind, as she headed north in the Mona Passage through the Antilles. A coupling on the engine's blower motor failed and one of the boilers had to be shut down while it was repaired. But from then until she reached Cape Hatteras, late on 12 December, the voyage settled down into an uneventful routine.

The Cape is marked by the Diamond Shoals light. From a position nine miles offshore, the *Argo Merchant* shaped a true, as opposed to magnetic, compass course of 040 degrees, straight for the Nantucket light vessel 400 miles away. The ship was equipped with both gyro and magnetic compasses and it was the relationship between these two over the next three days that was the key to the subsequent disaster.

The gyro was known to have an error of between one and two degrees so that, for example, to achieve a true course of 040 degrees, the helmsman had to steer 041 or 042 degrees. But this was no problem. The error was frequently checked and the Greek officers were happy to rely on the gyro repeater on the bridge for both automatic and manual steering. So much so, according to the subsequent Liberian investigation, that Second Officer Dendrinos had never had any experience of steering by magnetic compass. He would soon have to learn.

On 13 December the weather freshened from the north west and the tanker eased down to 7½ knots. By noon she seemed to have been driven several miles east of her projected course, even though the helmsman had been steering 036 degrees since four that morning to allow for some leeway. But twenty-four hours later a noon sextant sight indicated the opposite. The ship was now west, that is inshore, of the course she was trying to follow. Had a line been drawn through those two noon positions for 13 and 14 December it would have led almost exactly to the place where she ran aground by Fishing Rip. But the course was not changed, either then or later that afternoon when the wind backed south westerly, because Dendrinos reckoned – wrongly – that the tanker was being pushed back by an easterly current.

Just before 18.00 Chief Officer Ypsilantis returned to the bridge from supper and noticed that she was yawing from side to side more than usual under automatic steering, even allowing for the following sea. A check showed that the gyro on which they had all been relying was swinging rapidly back and forth through about six or seven degrees. The master switched out the automatic helm and ordered manual steering using the magnetic compass. The course he gave the helmsman was 047 degrees –

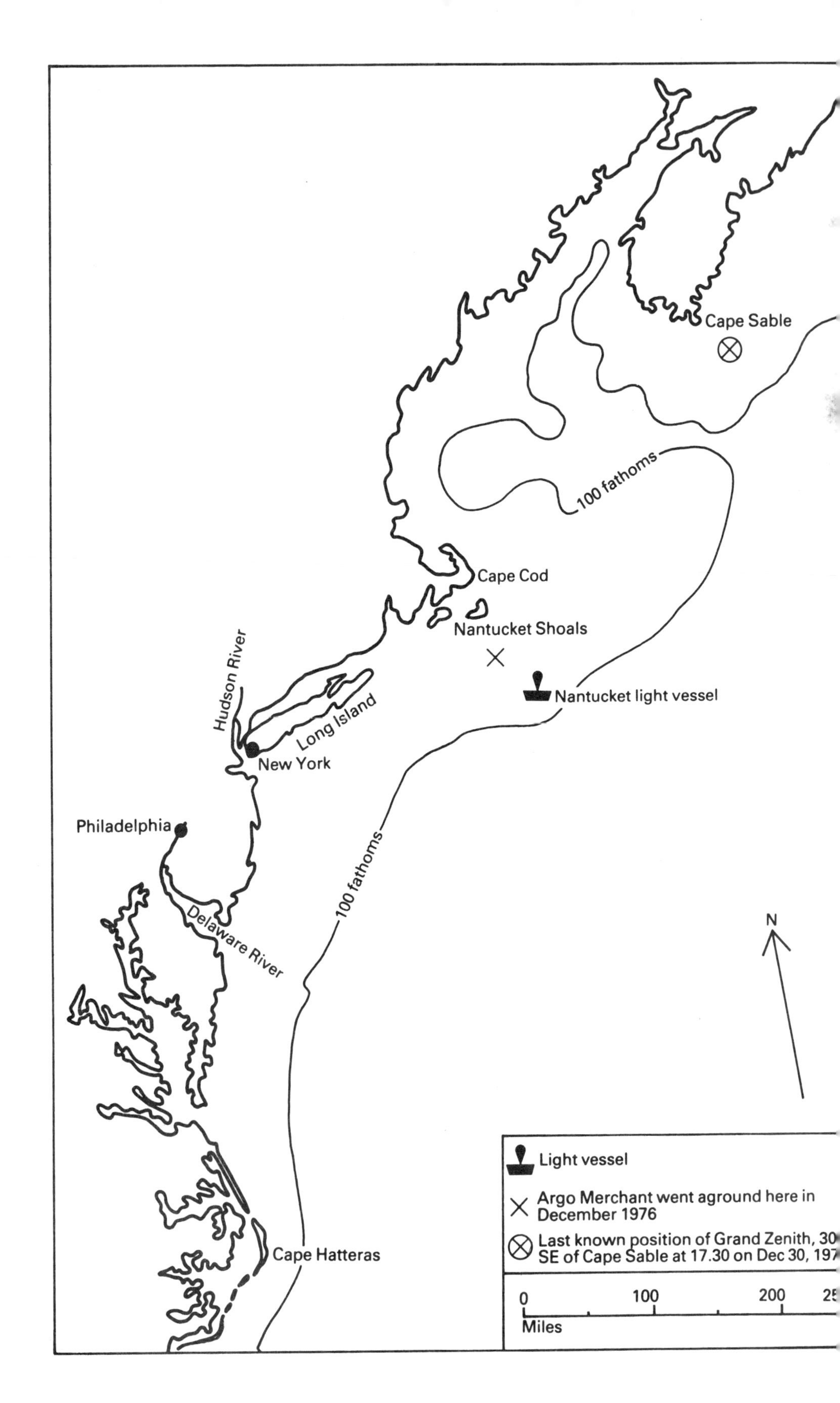
Cape Sable
100 fathoms
Cape Cod
Nantucket Shoals
Nantucket light vessel
Hudson River
Long Island
New York
Philadelphia
100 fathoms
Delaware River
N
Cape Hatteras
Light vessel
Argo Merchant went aground here in December 1976
Last known position of Grand Zenith, 30
SE of Cape Sable at 17.30 on Dec 30, 197
0
100
200
Miles

which is what the magnetic compass had been showing earlier that day when the tanker was supposedly on a true course of 036 degrees.

There were two reasons, quite apart from doubts as to whether the already known error in the gyro had been consistently allowed for, why this was a dubious procedure. As the vessel moved northwards from Cape Hatteras to Nantucket, the difference between true and magnetic North gradually increased from eight to fifteen degrees (a magnetic compass needle does not point to the geographical North Pole) and this was shown on the *Argo Merchant*'s charts. In her log, however, the difference between the two readings appeared to fluctuate and decrease as she moved northwards. With hindsight, therefore, it looks as if the gyro had begun to waver long before the evening of 14 December.

In any case Captain Papadopoulos would have done better – or so the subsequent Liberian board of investigation argued in its report – to decide first what true course he wanted to make good, and then add to it the westerly magnetic variation indicated for that bit of sea, which was fourteen degrees. By steering 047 degrees magnetic that evening, he was actually achieving a true heading of 033 degrees, not 036 degrees. However this discrepancy was later rationalised by talk of currents and so on, it was taking the *Argo Merchant* steadily inshore towards the Nantucket shoals, whereas the master maintained he was intending to clear the Nantucket light vessel by four miles to seaward.

Meanwhile the Chief Officer had hopefully transferred the basic course line the navigators were trying to make good from a small scale to a large scale chart of that area and 'walked' his dividers along it to mark the position the tanker should have reached at a given speed of about nine knots. When Second Officer Dendrinos came on watch at midnight on the 14th, he noted this 'dead reckoning' position and used it, with all its dubious assumptions, as the starting point for his own reckoning. With the magnetic course at last changed to 050 degrees, presumably in recognition of the increased variation from true North, he calculated that he should be abeam the Nantucket light vessel before the end of his watch, between 03.30 and 04.00.

At about the time he predicted, both he and the master saw an echo on the radar screen about five miles to starboard. But since it was on the 'wrong' side, they both apparently believed it was merely a tanker going the other way. Of the light vessel they could see no sign.

Left America's interest in tanker safety was spurred by a series of wrecks during the winter of 1976–7. Two of the worst occurred off the hazardous and crowded eastern seaboard.

When the Chief Officer came back on watch at 04.00 he naturally wanted to know where the light vessel was. He began to have an uneasy feeling that the ship had lost her way and urged the master – or so he claimed later – to 'do something' about it, by which he meant turn her round and go back. But at 04.30 both men were reassured, quite falsely, by switching on the radio direction finder and hearing the Nantucket signal apparently right ahead. So the elderly tanker went thumping on into the darkness without changing course.

Another hour passed without sighting the light vessel and the Chief Officer attempted, almost in desperation, to fix his position by sextant, using the Pole Star to determine his latitude and Capella to give him a crossing position line. Plotted on the chart, the position looked altogether unlikely and he assumed that the poor horizon had made his sights inaccurate. In fact, he had made a mathematical error. If he had done the calculation correctly, it would have shown the *Argo Merchant* about ten miles north west of the Nantucket light vessel, not so far from where she actually was and in evident danger of putting herself ashore.

In the event, Captain Papadopoulos seems to have shrugged his shoulders and suggested trying for a better star sight at 06.00. That was the exact time at which the *Argo Merchant* hit the bottom near Fishing Rip.

The Greek master immediately ordered the engines astern, although his Chief Officer protested that this would serve only to draw down the stern and grind the vessel deeper into the sand and rocks. At any rate she would not come clear and at 06.38 the US Coastguard was informed by radio. The wretched Ypsilantis had to make the humiliating admission that his ship was just somewhere within a thirty mile square north and west of the Nantucket light vessel.

There followed a long morning of indecisive radio and telephone calls, with the owners ashore still apparently trying to assess the situation and the Coastguard pressing for immediate action – a diving survey, tugs, lightening barges and so on. At 09.30 the Greek master had asked permission to jettison his cargo. But that was inevitably refused, and nothing more positive seems to have happened until 18.30 that evening, when Coastguard Commander Calicchio boarded the tanker. He found Captain Papadopoulos already in his shore-going clothes, a business suit and an overcoat.

Calicchio's Coastguard team were eager to get to work on some sort of salvage effort, but they seem to have received remarkably little co-operation from the tanker's crew. According to the Commander, half the crew had simply disappeared below. When he asked for a piping diagram of the ship, the master opened a suitcase in which he had already packed various official records and documents. The Chief Engineer seemed unaccountably vague about why the engine room was flooded with a mixture of oil and sea water, suggesting that a sea valve had probably been broken while the master was trying to back his ship off. The engine room was flooded to a depth of thirty feet. By lowering pumps into it, the Coastguard men were able to lower the level, but this was all they achieved.

At seven o'clock the next morning two salvage masters and a port engineer acting for the owners did turn up in a Coastguard helicopter. Yet even now they seemed to have no clear plan of action, and the port engineer did not help matters by promptly going to sleep. Commander Calicchio went ashore later that morning. By the following day the weather had worsened and there was nothing much anyone could do to save the tanker.

From then on it became a question of where the 28,000 tons of heavy fuel oil, intended for Salem, would actually end up. Fishing Rip is thirty miles off Nantucket Island. Cape Cod was about fifty miles away, Martha's Vineyard sixty miles. The *Argo Merchant* finally broke her back on 31 December and by good fortune winds carried the oil slick out to sea so that only the birds and the fish suffered.

The curious circumstances of the Greek tanker's loss, combined with the knowledge that ship and cargo were fully insured, naturally aroused public speculation as to whether she might have been driven ashore deliberately. It has certainly happened before. Marine underwriters joke grimly about a peculiarly deep corner of the Mediterranean where ageing, well-insured vessels often seem to founder. But the Liberian board of investigation emphasised that there was no evidence of malpractice whatsoever – just bad navigation and a suspicion that the Master might secretly have been pleased to find himself cutting the corner across the shoals so as to save his tide at Salem. The board's report summed up its assessment in this way:

> With the charter market bad, the possible stranding of the old vessel for her insurance has received public discussion. Unfortunately many of the circumstances are consistent with that possibility: stranding at high tide, reversing the engines to settle the vessel on the rocks over the protests of the First Officer, following a course for forty-eight hours directly to the point of stranding, no lookout on the bow, not heeding the fathometer readings, the Master's failure to save the compass error book notwithstanding the malfunction of the compass. . . . Yet with no direct evidence of such a plan or intent the board is loath to draw such a conclusion. It would presuppose a willingness to become entangled in an almost certain controversy resulting from damage to the environment, a controversy avoided only by the fortuitous drift of the vessel's cargo out to sea.

Controversy there certainly was, and an immediate response from President Carter. In his Congressional address on 17 March 1977, he proposed domestic measures to respond to a 100,000 ton oil spill within six hours, better provision for compensating those affected by oil pollution, and an international initiative 'to reduce oil pollution caused by tanker accident and by routine operational discharges from all vessels'.

To achieve these last two objectives he recommended US ratification of the 1973 Marine Pollution Convention (MARPOL) and a whole series of new rules for tanker design and operation. All tankers were to have segregated water ballast tanks to reduce operational pollution and a gas inerting system for empty cargo tanks, using boiler flue gases rich in nitrogen. They would also have improved emergency steering systems. Newly built tankers would have all these safety features plus double bottoms, to prevent leakage if they grounded without breaking up, and duplicated radar fitted with a collision avoidance system. The complete package of improvements was to be introduced within five years, at any rate on tankers wanting to use American ports, and the US administration made it clear that if the rest of the maritime world – that is IMCO – felt unable to move that fast, it would simply legislate unilaterally to protect its own shores.

In fact everyone agreed that nearly all the President's proposals were desirable in principle. The British Government, for example, had long been an advocate of inert gas systems and had begun to apply an IMCO resolution covering big, newly built vessels two years previously. But whether all Carter's ideas were practicable was a different matter. There was a suspicion on the British side, for instance, that the American computer lobby had hustled him into proposing elaborate collision avoidance radar whose operation and maintenance required great skill, when some mariners were already putting too much blind faith in the simple radars they had.

It was also widely agreed that IMCO must do something to speed up the process of ratification and enforcement. It was always drafting sensible rules and worthy resolutions that took many years to implement because insufficient members ratified them. The 1973 MARPOL, for example, could not come into effect internationally until fifteen states controlling at least fifty per cent of the world's gross tonnage had endorsed it – and that point had not been reached in 1978.

IMCO's answer was to start setting target dates by which its various measures should become effective. But some of its members, Britain prominent amongst them, nevertheless argued strongly that the United States would be wise to show more patience. American standards might be high, but if some other countries were encouraged to write their own rules, the whole delicate fabric of international controls could be destroyed. The better procedure was to establish a minimum international standard through IMCO first, amd then if necessary improve on it by national legislation.

The arguments came to a head at the Tanker Safety and Pollution Conference held in London in February 1978. The result was a compromise, both on the issue of unilateralism and between the powerful industrial lobbies whose interests were at stake. From their point of view, the most contentious proposal was that segregated water ballast tanks – to cut down on the amount of oily water pumped overboard – should be fitted in existing tankers as well as new ones. Countries like Norway and Greece with large, independently owned tanker fleets suffering from a desperately depressed oil transport market backed the American plan because it promised to send a lot of surplus tonnage to the scrapyard and reduce the carrying capacity of what remained by perhaps fifteen per cent. The British Government, on the other hand, was advised by its shipowners that the modifications would cost them something like £150 millions and increase the country's annual oil import bill by a similar amount; beside which they reckoned that a system of 'crude oil washing' developed by the major oil companies was just as effective in limiting pollution. They supported separate ballast tanks in new ships, as an alternative to double bottoms, because they could be placed round the hull to lessen the risk of oil leakage and fire in a collision.

In the end segregated ballast and crude oil washing were declared optional for existing ships and it was agreed that inert gas systems should only be phased in gradually once the relevant conventions came into force. IMCO committees were also briefed to take a closer look at duplicated steering gear in the light of the *Amoco Cadiz* failure and to prepare rules for collision avoidance radars. The 1974 Safety of Life at Sea (SOLAS) convention – not then in force – was modified to require annual safety surveys for tankers more then ten years old. Target dates were set for both MARPOL and SOLAS, which the US Coastguard promised to enforce whatever anyone else did.

The British and American delegations came out of the conference at the Cunard Hotel declaring themselves entirely satisfied with their compromise. The Norwegians went off gloomily predicting bankruptcy among their shipowners and probably feeling they had been sold down the IMCO river by their big American brothers.

3/The Sea in Danger

SHAKEN BY THE *Torrey Canyon* disaster, scientists, oil companies and governments have applied a great deal of effort since 1967 to finding ways of removing spilled oil from ocean and coast, both to protect the marine environment and, where the oil has come ashore, to protect the beauty or amenity value of the coastline.

Techniques include the use of floating barges to collect oil or protect sensitive areas; ingenious mechanical devices working on a variety of principles, including absorption into a porous material and the skimming and separation of oil and water; pumps and separators; absorbents spread on the oil, including straw, plaster and rubber powder, which is then collected in again; precipitants such as chalk, which mix with oil and sink it; and most controversially, but most widely used, chemical dispersants.

Dispersants have indeed become the most controversial subject in marine ecology over the past decade. Nothing causes more dispute among the experts. Sprayed on to the oil, the job of a dispersant is to use surface active agents and solvents to turn it into a stable oil-in-water emulsion of fine droplets that the natural processes of the sea can handle.

Developed in the early 1950's from an industrial paint brush cleaner, and used to clean oil from holiday beaches, the first versions were based on highly aromatic hydrocarbon solvents and non-biodegradable emulsifiers; and tests showed them to be highly toxic to marine life. Exposed to them, as many as fifty per cent of marine animals, particularly shellfish, died within forty-eight hours and British government scientists gave strong warnings about how and when they should be used.

But before the results of the toxicity tests were fully known, came the *Torrey Canyon.* With the wreck only fifteen miles from one of the most beautiful stretches of the English coastline, local and central government officials agreed that the economic impact of oiled amenity beaches was a greater danger than the threat dispersants posed to marine life. And as 14,000 tons of oil out of the 117,000 tons in the tanker drifted towards Cornwall, thousands upon thousands of gallons of dispersant were brought in.

It was clear that the developers of the dispersant technique had never intended it to be used on this scale, but with the army called in to assist, things evidently got out of hand. Dr Alan J. Southward of the Marine Biological Association of the UK, base at Plymouth, reported that 'In the emergency, few people were informed of the toxicity of the dispersant to all forms of life and indeed some of the spraying was carried out without enough regard to the safety of the operators themselves.'

Helicopters brought the supplies to cliff top dumps and, where the dispersant was spilled, the grass died. According to Southward, far from the dispersants being sprayed under the conditions recommended and pleaded for by biologists on the scene, more brutal techniques were often employed with 'opened drums simply rolled over the cliff'.

In all, 10,000 tons of dispersant was dumped on to the 14,000 tons of stranded oil. So toxic was the dispersant, Southward reported, that fifty per cent of the marine organisms exposed to it died within twenty-four hours at dilutions as low as five parts per million. Close to dispersant spraying, practically all animal life was killed. And in a report in 1978 on the ten years following the incident, Southward wrote, 'By itself the oil was not very toxic, although it killed some limpets and barnacles and most of the mortalities that followed cleaning were due to the dispersants.' And he added: 'If the wreck had happened on a more embayed section of the coast, or in an estuary complex, if the tanker had been carrying distilled products or one of the more toxic crude oils, if all the oil had come ashore and been treated with an equivalent ratio of dispersants, in these cases we might indeed have been closer to ecodisaster.'

Since 1967 optimistic reports have been written which suggest the marine environment in the affected areas of Cornwall has recovered rapidly, but, says Southward, this appears to be a myth which has somehow become enshrined in the literature. In fact, even ten years after 'it is the opinion of other ecologists besides myself that some shores are still not normal'.

The *Torrey Canyon* experience sent shock waves across the scientific world, particularly since other countries had stockpiled the same kind of chemicals, and as Southward said, 'It is perhaps the one tangible benefit of the *Torrey Canyon* that it drew universal attention to the dangers of dispersants.'

The French, who had watched in horror as 30,000 tonnes of the tanker's oil broke away and headed for the Brittany coastline, immediately rejected dispersants as a major solution and began to attack the oil with chalk, hoping to sink it. And approaches to other accidents following soon after, such as the Santa Barbara coastal well blow-out in California in 1969, and the wrecking of the tanker *Arrow* in Chedabucto Bay, Canada, in 1970, were drastically revised to place less emphasis on chemicals. At the same time the incident had the effect of spurring research into mechanical methods of oil recovery which until then had hardly existed.

Today, a new form of dispersant has been developed, using non-hydrocarbon ingredients and a biodegradable emulsifier. Made in a concentrated form to be mixed with water just before spraying, it has a claimed toxicity level a thousand times less than that used in 1967, and Britain is replacing all existing stocks with it. But not all first generation stock has disappeared from all countries, and thousands of gallons of a low aromatic second generation dispersant will still be on call for some time to come.

Ironically it has been the British who have led the research in this field and who have held on to dispersants as a major weapon against oil spills. Dr Douglas Cormack of the Government laboratories at Warren Springs, near Lowestoft, explained: 'Even though certain toxic effects had been noticed in beach areas, none had been identified at sea. Again, observed shoreline toxicity could be attributed to over use or misuse of dispersants. It was therefore decided to concentrate on dispersants as the most promising method and develop it.' The British took one other factor into consideration, their geographical position; the waters of the North Sea, the Atlantic, the English Channel and the Irish Sea which surrounded the country were known for a predominance of rough conditions, the kind of situation in which other oil removal methods would be hampered or useless. The Ministry of Agriculture Fisheries and Food concluded: 'Dispersants are only used where harm may already be occurring due to the presence of a spill. Indeed, in the absence at the present time of any effective methods to recover or contain a slick under normal North Sea conditions, dispersion by chemical means may be the only way of protecting seabirds, inshore shellfish resources, or beaches of high ecological, fishery or amenity value.'

Most marine scientists have now moved towards a position where they are prepared for the use of the new dispersants in the open sea, to prevent oil coming ashore. But not all of them are happy about the use of chemicals close inshore, or near shellfish beds or breeding grounds. The major problem, twelve years on, remains exactly that of the *Torrey Canyon*: the over use or misuse of the product.

In action, the dispersant is designed to be sprayed from a moving boat, behind which are towed boards to agitate the water and mix oil, dispersant and sea together. But in the face of the emergency conditions of an oil catastrophe many things can happen: not enough spraying equipment can be found; the dispersant is not mixed correctly before spraying; boat skippers go too fast or cannot be bothered to tow the agitator boards; badly controlled spraying sends most dispersant into clear water; or dispersant is used where wind and tides can carry it on to environmentally sensitive shores. Unless the dispersant is itself thoroughly dispersed it cannot act upon the oil: incompletely treated oil simply re-coalesces later in a different place. The temptation too is to see dispersant as a panacea for all ills, with boats being sent out to spray regardless of conditions or of the oil being treated, so that authorities can be seen to be doing something. In May of 1978 up to nineteen vessels sprayed heavy fuel oil spilled from the *Eleni* V in a particularly ineffective exercise. The oil, which had been carried in heated tanks, solidified in the water into a form no dispersant could treat.

And not all scientists are yet convinced that even the latest dispersants do not hold dangers. Dr Southward is forthright: 'Let us make no mistake, all dispersants are toxic to some extent, even the water-based mixtures that kill at one part per thousand by themselves. Experiments show that these so-called non-toxic dispersants can increase the lethality of oil. One of them, mixed with an equal part of Iranian crude can kill or prevent the normal development of the eggs of marine fish at dilutions of only five parts per million. This mixture is therefore as toxic as the older generations of solvent based dispersants used in Cornwall in 1967, thousands of drums of which are still stockpiled round the country.' A group of United Nations experts who looked at the problem for the Food and Agriculture Organisation, agreed and disagreed. 'The concern and conclusion that all chemical dispersants are in themselves inherently toxic is incorrect,' they concluded. But added 'The toxicity of the dispersed oil itself is a more valid concern.' Experiments discovered that while some oils were exposed to quicker biodegradation by dispersants, 'emulsifying the oil increased the leakage of soluble components of the oil into the sea water'. The overall effect of the dispersion was to introduce fine droplets of dispersed oil several feet or more down into the water where it would be more susceptible to biodegradation; and the finer droplets the dispersant could produce the more efficient it was considered because the more stable was the emulsion produced. But the finer the droplets the easier it was for the oil to be taken up by marine life. 'Probably of greater concern than the acute effects is the fact that the finely dispersed oil droplets become a more widespread contaminant and may cause long term effect,' the experts reported. And even the oil industry's own team of pollution advisers the International Tanker Owners Pollution Federation which believes in the careful use of dispersants as a practical weapon, told a conference in Los Angeles in March 1979 that 'on many occasions leaving the oil to disperse and degrade naturally will remain the best response'.

Just how serious the continuing injection of oil into the marine environment may be, both for the creatures which live in and around the ocean, and for man himself, is the subject of unending debate and expanding research by the world's marine scientists, but the essential truth remains that no one yet has enough knowledge to pretend to know all the answers.

The consensus for the moment is that lasting ecological damage to the high seas has not been verified, and that oil pollution should be treated as an inescapable fact of life; something which has increased over the years but which has now reached a plateau, or is even beginning to decline. And that while it remains an important and legitimate field for scientific study, it is well past the stage where the emotional furore stirred up by the conservationists after each tanker disaster needs to be given serious consideration. Those who believe this summarise their case as follows:

Most surface and near-surface waters of the ocean contain hydrocarbons in a range between one and ten parts per billion. In deeper waters the concentrations are smaller, often less than one part per billion. This equals one drop of oil in about 26,000 gallons of water; and at such concentrations there is believed to be little or no toxic effect on marine life.

So the only major damage by oil pollution of the sea, is to the marine amenities increasingly enjoyed by man. Direct damage to shores, the killing of seabirds, and the deaths of fish, marine animals and plants are limited and localised; more certain damage caused by chronic pollution near refineries, oil terminals and petrochemical plants can be cleaned up, at a price; and the scare-mongering fears of the conservationists, about the possible passage of carcinogens in the oil up the food chain to man, remain unproven.

Some analysts go further. The world oil pollution crisis peak may even have passed, they argue. The end of oil's dominance as a source of energy and chemicals is within sight and may well occur within the lifetime of some of us. In recent years, big oil price rises by producer countries and the need to explore new and vastly more expensive fields, such as the North Sea, mean that oil has changed from being a cheap raw material costing hardly more than the price of transport from well to refinery, to a valuable and diminishing resource providing the incentive to prevent wasteful spills.

At the same time, public feeling about the environment coupled with stricter international legislation has forced oil companies to tighten or re-arrange their procedures so that, it is said, less oil is finding its way into the sea today than there was twenty years ago – despite an increase in world consumption and the larger amounts being carried across the oceans.

Not all scientists are so confident. They look out over the ocean from the polluted beaches and are concerned that even the controlled levels of oil pollution today may already be creating unknown and far reaching effects as the years go by; particularly since oil is only one of a number of pollutants entering the seas: others include domestic sewage, pesticides, inorganic wastes and heavy metals, and radioactive materials. They admit their lack of knowledge and press for more urgent work to be done to determine just what these effects may be, in case we are now approaching a time when the sea may no longer be able to handle what is dumped into it. The state of enclosed seas like the Mediterranean and the Baltic already gives them cause for concern.

It was the Swiss oceanographer Jacques Picard in 1972 who warned that if pollution continued, life in the sea would be extinguished within twenty-five to thirty years. In 1978 the Norwegian explorer Thor Heyerdahl added his view: 'There are people who tell you that oil doesn't matter, that the sea can absorb and recycle all this pollution. I call them the Sandmen: they want to put you to sleep with calming words. Don't listen! Unless you and I – all of us – act now to stop the seas being overloaded with poisonous refuse, they will suffocate and die.'

Those who feel this way do not accept that anything about oil is a constant. Despite the evidence that oil will have to become less important as a fuel, they are not convinced that man's increasing numbers and his appetite, particularly as the developing countries become more advanced, will allow this to happen. The known world reserves already run into hundreds of billions of tonnes and new sources are being found constantly: Mexico, once written off as a supplier of minor consequence has now been shown to have reserves which may make her the sixth largest oil producer in the world. There is still a vast amount of oil in the ground.

And, they insist, the real parameters for oil pollution can be drawn much more closely. The vastness of the ocean is misleading: the important part of the sea for the production of marine life of all kinds (and thus perhaps the nursery of the ecosystem) is the coastal zone above the continental shelves. The organic production of this zone has been estimated as fifty times that of the open sea, and yet it covers only 7.6 per cent of the ocean surface.

The coastal areas are also, of course, the interface between man, the oil industry and the natural environment: the sites for the refineries, the petrochemical plants and the oil terminals; the destination for the supertankers and the almost inevitable scene of the regularly recurring catastrophes involving tankers. They are the site of the arrival of the river

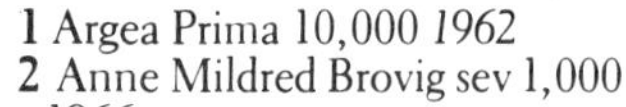

1 Argea Prima 10,000 *1962*
2 Anne Mildred Brovig sev 1,000 *1966*
3 Torrey Canyon 100,000 *1967*
4 Ocean Eagle 45,000 *1968*
5 World Glory 45,000 *1968*
6 Andron 16,000 *1968*
7 Keo c.25,000 *1969*
8 Pacocean 30,000 *1969*
9 Albacruz 20,400 *1970*
10 Sofia P. 18,620 *1970*
11 Gezina Brovig 16,000 *1970*
12 Arrow 12,000 *1970*
13 J. L. Anastasia 18,500 *1970*
14 Polycommander 16,000 *1970*
15 Ennerdale 40,000 *1970*
16 Silver Ocean 18,300 *1970*
17 Pacific Glory 6,300 *1970*
18 Marlena 15,000 *1970*

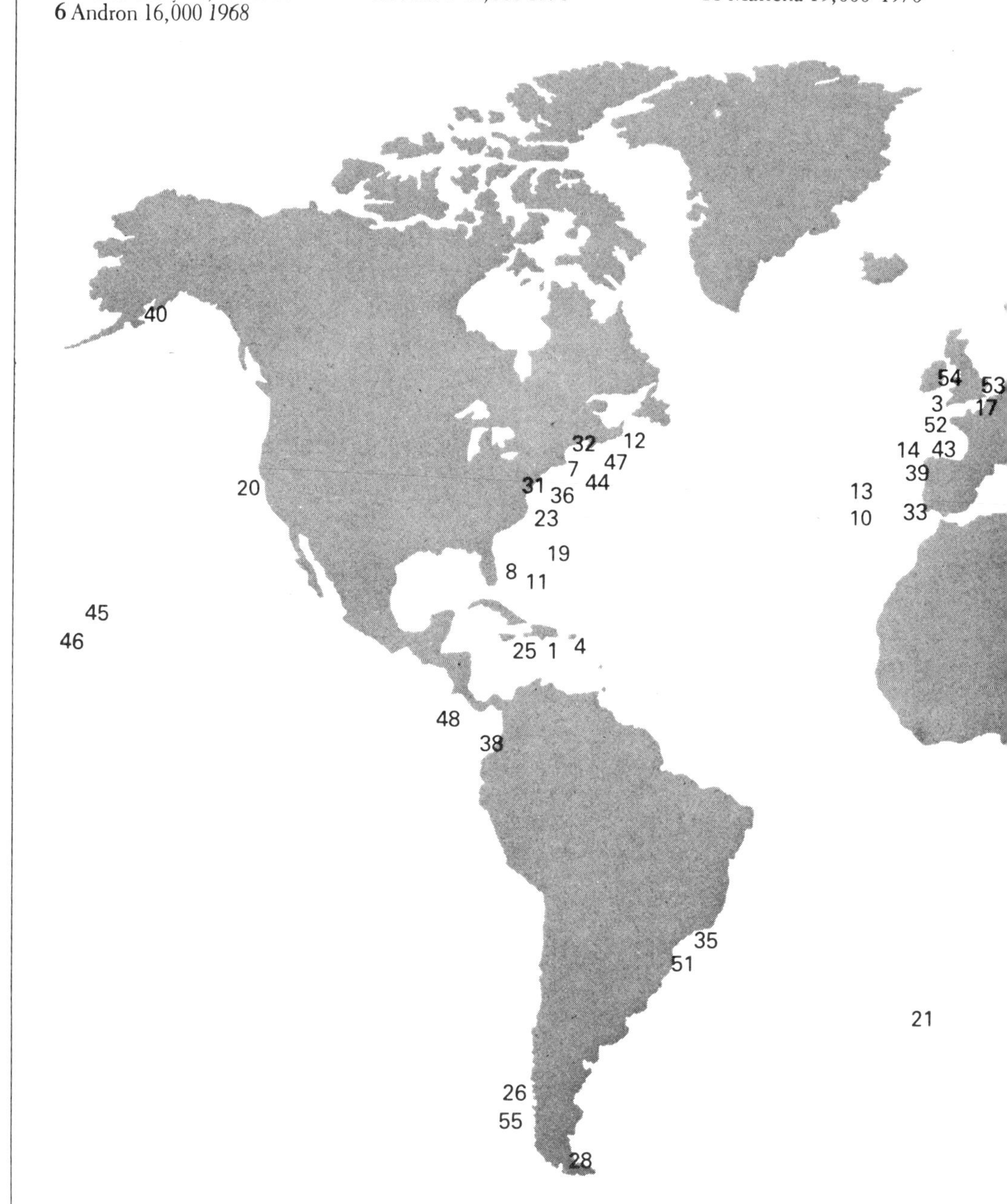

19 Chryssi 31,000 *1970*
20 Oregon Standard 7,750 *1971*
21 Alkis 18,000 *1971*
22 Wafra 64,000 *1971*
23 Texaco Oklahoma 31,500 *1971*
24 Sea Star 65,000 *1972*
25 Zoe Colocotroni 8,000 *1973*
26 Napier 30,000 *1973*
27 Jawacta 20,000 *1973*
28 Metula 56,000 *1974*
29 Trans Huron 5,000 *1974*
30 British Ambassador 50,000 *1975*
31 Corinthos 5,000 *1975*
32 Athenian Star 10,000 *1975*
33 Jakob Maersk c.8,000 *1975*
34 Scapbay 13,000 *1975*
35 Tarik Ibn Ziyad 12,000 *1975*
36 Spartan Lady 25,000 *1975*

37 Nanyang 20,000 *1976*
38 St Peter c. 30,000 *1976*
39 Urquiola 18,000 *1976*
40 Sealift Pacific 5,000 *1976*
41 Cretan Star 28,500 *1976*
42 Diego Silang 5–6,000 *1976*
43 Bohlen 11,000 *1976*
44 Argo Merchant 32,000 *1976*
45 Irenes Challenge 36,000 *1977*
46 Hawaiian Patriot 17,500 *1977*
47 Grand Zenith 24,000 *1977*
48 Caribbean Sea 35,000 *1977*
49 Al-Rawdatain 7,000 *1977*
50 Venoil/Venpet 15,000 *1977*
51 Brazilian Marina 17,000 *1978*
52 Amoco Cadiz 230,000 *1978*
53 Eleni V 5,000 *1978*
54 Christos Biṭas 6,000 *1978*
55 Cabo Tamar 6,000+ *1978*

Oil spills from tankers, 1962–78. Only spills exceeding 5,000 tons are shown. Note the concentrations on the eastern seaboard of the US and in the approaches to the great ports of Western Europe. (Plus key attached).

and urban run-off (seen by the National Academy of Sciences as the biggest single source of man-made pollution) and, through rain and river, the arrival of atmospheric hydrocarbon pollution, the discharges of motor cars, power stations, industrial furnaces and domestic central heating Even in Exxon's research, claiming there was little or no effect on marine life, readings in coastal waters showed up levels of 100 to 200 parts per billion.

Two types of pollution here cause concern. The first is the chronic pollution, the continuing high level of hydrocarbons in the sea from all sources, from refinery effluent to tanker discharges, which by being present year after year may be establishing imperfectly understood long term effects. Already, as an example, some ornithologists say oil pollution is limiting the numbers of seabirds in areas where oil is a recurring problem: long-tailed ducks in the Baltic, guillemots, razorbills and puffins in the English Channel and in the Gulf of St Lawrence, have all been reduced. A close watch is being kept on estuaries, saltmarshes and foreshores near oil installations to record changes in plant and animal life to see what lives and what dies and to try to assess its significance to the environment.

The second is the single exceptional spill from a disaster such as that which overtook the *Amoco Cadiz*. No single spill has ever been as large before but several have been big enough to alarm the conservationists and the more doubtful ecologists. A big coastal spill has immediate consequences for wildlife and for man: disrupting the ecology of an area, often with a huge if localised mortality of species; wiping out the amenity value for some time of an attractive stretch of coastline; and economic damage to those who earn their living from it. But it may also have more far reaching effects.

The oil can be cleaned up, and the visitor, arriving only a few months later, may conclude that all is well once more. But the balance of the local ecosystem has been badly disturbed, and as it tries to regain equilibrium over several years it can be seen as swinging like a pendulum, first to this combination of plant and animal life, and then to that, until the swing slows and the system returns to its previous balance. Cornwall after the *Torrey Canyon* was an example.

The danger of such spills if repeated, say some scientists, and given the background of the already present chronic pollution, is that the pendulum may swing one way and never come back: that large sections of the ocean may stabilise at a new and different level with the mixture and diversity of their marine life completely changed.

Apart from more tangible results, such as, perhaps, the disappearance of locally important species of commercial fish, the scientists point to a much wider and more subtle potential effect. People come to Atlantic coast areas like Cornwall, Brittany and Maine for their 'atmosphere', a combination of weather, geology and, importantly, the marine life of all kinds which marks them apart from other types of coastline. Ecological damage from oil pollution could affect enough species to change that atmosphere, indeed change the atmosphere of any individual coastline at some date in the future until they are all very similar: with no seabirds other than the ubiquitous gulls and covered with the same green weeds. Very much the same and uninteresting. This possibility alone places an awesome responsibility on governments, and the oil companies, to do all they can to establish an even firmer grip on the problem of pollution.

And not all scientists can live with the individual findings of the consensus, which insists man is not at risk. Southward of the Marine Biological Association says: 'Oil degraded . . . by natural forces, or chemically by dispersants, is greatly diluted. In theory it presents little further risk to marine life. But if repeated spills occur along the same stretch of coast we cannot afford to be so sanguine. The hydrocarbons may accumulate to a level where there could be a risk of malignant tumours in fish and man. Such a prospect is by no means imaginary, for a little publicised aspect of many oil spills is the quantity of oil that persists for years in some situations, for example buried in sand, forming a chronic source of pollution which can maintain excessive hydrocarbon levels in the local marine fauna. From available data it looks as if the *Amoco Cadiz* spill will be no exception in this respect.'

Nor do all authorities believe that the day to day contributions by man to the hydrocarbon levels of the sea, tanker catastrophes apart, can be accepted in comfort. Dr Claude Chassé of the University of Western Brittany says: 'The spectacular character of tanker accidents, due more to negligence and greediness than to the hazards of incidents at sea, should not mask the pollution which is more insidious and more important in volume, from the tank washing at sea and the spills of urban and industrial effluents which lead to hydrocarbon levels often near to 0.1 parts per million which render the products of the sea uneatable: this threshold would probably have been attained if the oil crisis had not intervened.'

4/Crowded Waters

EARLY IN FEBRUARY 1978, the *Amoco Cadiz* slipped her moorings in the Persian Gulf and set sail for Holland. Although she was effectively American-owned, she was flying the Liberian flag of convenience. Deep laden with 223,000 tons of crude oil for the refineries in Rotterdam, she moved away slowly but steadily, gradually working up towards her fifteen knot service speed.

Her Italian crew were looking for an easy trip and they had every reason to expect one. Their ship was only five years old, with all the latest navigational aids and, for a VLCC like her, the Cape run was entirely routine. And so it proved for most of the voyage. But as the giant tanker left the open waters of the Bay of Biscay for the crowded shipping lanes of the Channel, things began to go horribly wrong.

The weather had already deteriorated. Even the great bulk of the *Amoco Cadiz* was pitching and yawing in the south westerly swell built up over the edge of the continental shelf by a rising gale – not yet the most dangerous kind of onshore gale, but promising to veer north westerly later. It was a situation which had brought disaster to many ships in the past – for the *Amoco Cadiz* was making a landfall which had been welcomed, and feared, for centuries.

We'll rant and we'll roar like true British sailors,
We'll rant and we'll roar o'er all the world's seas,
Until we strike soundings in the Channel of old England,
From Ushant to Scilly is thirty five leagues.

The words of the shanty refer to the 'gateway' through which most North Atlantic shipping must pass to reach the great ports of North West Europe. *Torrey Canyon* hit one gatepost, *Amoco Cadiz* the other.

In the days of the square rigged clippers, sailors knew they had reached the relatively shallow water of the Channel because they were able to touch, or sound, the bottom with a lead line. This was a long line marked in fathoms with strips of leather, linen and bunting – so they could be identified by feel as well as sight – and weighted with a lump of lead. The bottom of the lead was hollowed out and 'armed' with tallow to pick up a trace of the sea-bed. Hence the hydrographers' enthusiasm for marking charts with little symbols – *s* for sand, *sh* for shell, *m* for mud, and so on – to indicate the nature of the bottom. And this was no theoretical exercise. There must be many tales like that of the old skipper who was sounding his way into Waterford on the southern coast of Ireland, at night, with a south westerly gale threatening to sink his sixty-year-old trading schooner under him, and who put his helm up to run blindly into the strange port when the mate's lead found mud instead of rock and sand. It meant he had found the channel.

Nowadays ships have echo sounders, radar, radio direction finders, even space satellite navigation equipment; but they still come to grief in the Channel's bottleneck, choked at its narrowest point by the notorious Goodwin Sands, and little deeper in some places than the twenty metre draught of a supertanker.

A loaded 200,000 tonner in-bound from the Atlantic would normally hold to the southern side of the Channel along the French coast. To begin with she has a fair freedom of manoeuvre – provided she can manage to avoid running into Ushant and the Cherbourg peninsular. But as she approaches the narrow Dover Strait her path is blocked by a whole cluster of sandbanks. Some, like the Goodwins, have always been feared by mariners. At low tide the masts of wrecks are scattered like dead trees on the yellow sand. But others are really only obstacles to a vessel of the VLCC's enormous bulk and draught. The water available to her on the French side narrows to no more than about five miles between the Vergoyer and the Bassurelle, and again between Les Ridens and the banks off Boulogne. At the Sandettie, the maritime crossroads of the Dover Strait, the recommended deep water route for tankers bound for Rotterdam is only a mile or two wide.

This might seem a generous enough allowance for a professional navigator aboard a well-equipped ship. Yet the record shows that some of the navigators are professional in name only, and of course the best equipment can sometimes let them down. It was near the Sandettie light vessel in 1978 that the VLCC *Al Faiah* suffered a gyro compass failure that might have put her straight up on the South Falls bank, had the Coastguard not noticed her strange course on their radar and alerted her by radio.

In any case, a mile of sea is really not much to play with when you are driving a supertanker. A loaded 200,000 tonner pushing ahead at fifteen knots takes not one, but two and a half miles to stop, over a period of more than twenty minutes. The astern power is so inefficient it is scarcely worth applying until most of the way has been taken off her. In fact it is quicker, if there is enough room to port or starboard, simply to put the helm hard over and turn the ship through 180 degrees – when she will advance about half a mile before turning back and eventually achieve the reciprocal heading perhaps two thirds of a mile to one side or the other.

Hardly a dainty manoeuvre, but when your ship is built like an underwater aircraft hangar it is the best you can expect without expensive additions like twin screws and rudders. At low speeds manoeuvrability is even worse, and at less than about three knots a loaded VLCC is virtually unsteerable.

She still has to be steered, of course, with or without tugs, and the particularly awkward low speed manoeuvres often have to be made in shallow, confined waters or in a crowded port approach channel. There, the economic pressure to use bigger and bigger vessels without the expense of lightening them offshore may force a Master to attempt a channel with no more than three or four feet of water under his keel.

Again, this might possibly seem a comfortable enough margin until one sits down to draw the cross section of one of these giants to scale, with a flat bottom 150 feet wide and only a three foot layer of water beneath it. In practice she will have begun to 'sniff' the sea-bed before it gets that shallow, making an already clumsy creature still more sluggish. She will also be suffering from a curious maritime affliction known as 'squat'. This is the tendency of a moving ship to sink a little because of the depression of the water surging past the hull – a tendency that is accentuated in a confined channel and at higher speeds. A big tanker may squat four or five feet and, rather oddly, she tends to trim down by the head, so that her progress through a shallow waterway, almost nuzzling the bottom with her long, bulbous nose, neatly fits the other metaphor about sniffing.

When the VLCCs began to appear in the mid-1960's, Exxon decided that a radical training initiative was needed to acquaint its captains and harbour pilots with the handling characteristics of these giant creatures. The oil company got together with a French hydraulics firm and remodelled a lake in the foothills of the Alps, at Grenoble, to represent typical channels and harbours at a one-twenty-fifth scale. Trainees chugged round in scaled down model tankers, practising approaches, berthing, mooring to offshore terminals, all without risk to anything but their professional pride. They could even try their hand at the most awkward section of the Suez Canal.

Tanker operators also make increasing use of training simulators – one of the more obvious ways in which they now work in parallel with the aviation industry. In flying, ground simulators have long served the basic purpose of preventing trainee pilots paying for elementary mistakes with their lives. That requirement remains, but has been joined by an economic one, the need for a cheaper way of converting pilots from one type of airliner to another than by flying the machine around empty. At sea, applying the wrong compass variation or misjudging a tide are not usually fatal, but can cost an awful lot of money and pollution, so it is this second factor that has prompted the development.

The most basic ship simulator might consist of no more than the mock-up of a navigational bridge, but as with aircraft simulators, the real fascination of such machines lies in the attempt to represent what goes on outside the windows. Many of the modern flight simulators cope with the difficult landing and take-off phase by scanning a detailed model of the airport and the surrounding countryside with a camera whose image is projected on to a screen. This gives an extremely convincing realism unless you 'fly' off the model, when the camera falls back on an unhelpful shot of cotton wool clouds. However, scanning plasticine waves and model buoys solves only half the problem of a ship simulation because it is almost impossible to introduce other vessels moving and reacting to the trainee's manoeuvres.

The expensive solution is to use what is known as computer generated imagery. But a radar engineer called Pat Hansford, the epitome of the British boffin, working in a shed at the back of Decca's factory in Hersham, has produced a much simpler answer. His team's design was based on the realisation that if one could simulate the worst navigational case, the rest could reasonably be ignored. The worst case at sea is by night, when the navigator has to make sense of a moving perspective of red, white and green lights – yet this is far easier to simulate on a darkened screen, with the dim outline of a tanker's bows superimposed. Half an hour in such a simulator, with the 'engines' throbbing beneath the deck, and it takes a mariner of strong nerves to ignore that red navigation light appearing in the wheelhouse window to starboard.

Another new requirement in an age of supertankers is better charts, or at least more accurate surveys of areas that could previously be neglected. Until the appearance of the echo sounder in the 1930's, the depth of the sea-bed had to be painstakingly surveyed with a hand lead or a sounding machine (no more than a mechanised version of the leadline using a reel of wire and a boom). Today with accurate echo sounders and side-scanning sonar adapted from naval equipment, it is still a slow, tedious and expensive business. So even some of the famous British Admiralty charts have not been given a major revision this century, and only about a quarter of them have been substantially revised in the past ten years.

The Hydrographer of the Navy is responsible for them because until quite recently it was generally the naval requirement that set the standard of charting in a given part of the oceans. But the increasing size of merchant ships, first the passenger liners, then supertankers of up to half a million tons, has gradually changed that. Submarines do still make special demands on the naval hydrographers, but it is the deepening draught of the giant crude carriers that has forced chart makers to lower what they call the survey danger line – the depth below which the sea-bed need not be fully examined. And where the water depth is only marginal anyway, as in the eastern part of the Channel and the North Sea, wrecks have also had to be more comprehensively charted and surveys taken more frequently to check shifting banks and silting channels.

In just one year's work, in 1974, the Hydrographer's fleet discovered ninety-five uncharted wrecks in the Channel and a dangerous shoal in what everyone thought was deep water off the east coast of England. It was surprises like this, plus the fact that commercial demands on the hydrographic service worldwide were outstripping the ability of the British defence budget to pay for them, that led to the establishment of a government study group that same year. It had to admit that only twenty-four per cent of the UK's continental shelf was surveyed to full modern standards and that seventy-two per cent – 134,000 square miles – was either unsurveyed or surveyed only by leadline. Yet the group's report maintained that 'only full modern standards are sufficient to ensure safe navigation within five metres of the sea-bed'. To bring the whole of the continental shelf up to this standard, it was calculated, would take 284 survey ship-years of work.

Three years later, HMS *Fawn* was surveying a recommended shipping route to the Irish Sea when she discovered a 600-yard-wide sandbank at a depth of twenty-one metres (sixty-nine feet) where the chart indicated general depths of about fifty metres (165 feet). It turned out the area had not been investigated by a survey ship since 1843, when a sailing vessel used a hand leadline to take lines of soundings threequarters of a mile apart.

That same autumn, HMS *Bulldog* started checking out the wrecks on the English side of the Channel near the Varne sandbank, where no less than 169 had been discovered in 1960 using less sensitive sonar. During the intervening years, about sixty of them seemed to have disappeared, but another sixty were found, many of them with less than twenty-three metres (seventy-six feet) of water over them.

The peculiar hazard on the French shore of the Channel which claimed the *Amoco Cadiz* is the strength of the tides, and the fact that it becomes a lee shore in the prevailing westerly winds off the Atlantic. The tidal range is twenty to thirty feet with currents to match. The coastline itself is also of an especially uncompromising kind, jutting out in rocky headlands and islands at Ushant and the Casquets. Past the Cherbourg peninsular the shoreline softens but the tides are still strong and they remain so through the sandy narrows off Cap Gris Nez.

About 300 ships a day, that is more than 100,000 a year, are funnelled through the Dover Strait, and about fifty of them are tankers carrying oil, gas or chemicals for the industrial centres of North West Europe. Scuttling across their path during the summer months, the passenger ferries make about 250 crossings a day. It is little wonder that this is one of the world's marine accident black spots. To make the point more vividly for visiting politicians and journalists, the Coastguards who man the radar station overlooking the strait from Langdon Bay show a frightening film of the blips on their screen speeded up until they flow through the bottleneck like a stream of phosphorescent ants, with the echoes of the ferries bouncing back and forth between them. It looks as if collisions are sooner or later inevitable and of course they are, though on past evidence it is not the agile passenger ships that cause the trouble.

A recent British government study of marine casualties calculated that on average there is a collision or a stranding once in every 25,000 transits of the Dover Strait. Since the number of loaded oil tankers is about 8,000 a year, one must face the statistical probability of a casualty every three years, and perhaps every ten years a really big spill.

The same study went on to conclude that it would be unreasonable to provide the permanent dispersant and clean-up resources to handle a major disaster of this kind: 'The objections to a new policy of gearing up to cope with the "worst possible incident" are that this would entail tying up large resources to deal with a disaster which might occur once a decade; and that even this scaled up capability would not, in the worst case, prevent oil coming ashore in substantial quantities. Also the resulting pollution, though it is damaging to the environment and to amenities, and a matter of great public concern, does not cause loss of life.'

The most serious casualty that actually did occur in the Channel during the eleven year period between *Torrey Canyon* and *Amoco Cadiz* was the collision between two Liberian tankers, the *Pacific Glory* and the *Allegro*, in October 1970. Apart from the 6,000 tons of oil that was spilt off the Isle of Wight, it was remarkable for two things: the fact that the two vessels were within sight of one another, on an exceptionally clear night, for at least half an hour before they collided, and the remarkable way in which what seemed at first to be no more than a serious incident, with only minor collision damage, was transformed two hours later into real disaster, with a heavy loss of life.

The *Pacific Glory* was a 78,000 ton motor tanker, registered in Monrovia, crewed by Chinese, and on charter to Shell; the *Allegro* a steam tanker of 95,000 tons, also registered in Monrovia, but with a Greek crew and chartered to Esso. Loaded with crude oil, they both reached the Channel on 23 October after uneventful voyages from Nigeria and Libya respectively.

The *Allegro* came up the French coast from Ushant to the Casquets and then crossed over on a course of 050 degrees for St Catherine's lighthouse on the Isle of Wight. Her final destination was the Fawley refinery in Southampton Water. The *Pacific Glory*, on the other hand, entered the Channel on the English side so as to pick up a pilot to take her right through to Rotterdam. Captain Jochem Frudiger boarded her that afternoon off Berry Head and set a course of 087 degrees to take her straight to the Bassurelle lightship in the entrance to the Dover Strait. By about seven o'clock in the evening they were both near the Isle of Wight on slowly converging courses, the *Pacific Glory* intending to go straight past and the *Allegro* closing her from starboard at a marginally higher speed of 15½ knots, compared to the motor tanker's 15 knots. At 19.40 the *Allegro* altered course to 060 degrees, narrowing the angle of intersection to 27 degrees.

It was just about then that Second Officer Leung Yi Yeung aboard the *Pacific Glory* first noticed the lights of the *Allegro* to seaward. He saw the other tanker's red port side navigation light and her white masthead lights about a mile away and just abaft the beam. He judged that she was on a roughly parallel course and slowly overtaking his own ship.

This judgement – that the *Allegro* was overtaking – was absolutely crucial to what happened later. Under the IMCO rules for preventing collisions at sea, it is the overtaking ship's responsibility to keep clear of the one being overtaken. So if the *Pacific Glory*'s Second Officer was right in his assumption, which was shared by the Dutch pilot, the *Allegro* should undoubtedly have kept clear. But if the two tankers were considered simply to be crossing one another's path, then an entirely different rule applied, and the responsibility was the other way round.

Under that rule, the ship which had the other on her starboard side – the *Pacific Glory* under this interpretation – should have given way. And in that case the *Allegro* was not only entitled to hold her course and speed, she was bound to do so, in order to avoid confusion, until a collision could no longer be avoided by the *Pacific Glory*'s action alone (though to make things really complicated, this rule has since been amended to give the stand-on vessel earlier freedom of manoeuvre in precisely this kind of situation).

Aboard the *Allegro* that night, however, nobody was debating such technicalities. According to Captain Vassilios Vastardis, no one had even noticed the *Pacific Glory* until 20.17, when he himself suddenly saw her – the two white masthead lights, the green starboard side light and the accommodation lights aft – about a mile away to port, just forward of the beam. One of his seamen, known affectionately by the crew as 'the old man', said later that in fact he had been on the wing of the bridge since just after eight o'clock, and had seen the other tanker's lights, but when he called out 'vessel to port' no one inside seemed to take any notice. He was not formally posted as a lookout anyway.

What happened over the next five minutes we can never truly know, because the two crews' accounts are contradictory. It seems that just before the collision, the angle of intersection was narrowed still further by the *Pacific Glory*'s altering course slightly to port to avoid a third, still distant ship ahead. The *Allegro*'s Greek master was meanwhile feebly trying to alert the other vessel by flashing his masthead 'blinker' light, but the Dutch pilot evidently did not order hard-a-port until the *Allegro* was right on top of him. Almost simultaneously, Captain Vastardis gave orders to put the *Allegro*'s helm hard-a-starboard and stop the engines.

As the two big tankers turned away from each other their sterns swung together. The *Allegro* struck the *Pacific Glory* just about level with the bridge, buckling the steel plating of the aftermost cargo tank on the starboard side, the engine room, and the pump room and cofferdams that separate them. As soon as the two ships came clear, the *Pacific Glory*'s engines were also stopped and the pair of them slowly drifted to a halt in the darkness while their crews tried to sort out what had happened.

After the alarm of the past few minutes, the actual collision was almost an anti-climax. Captain Frudiger described it as no worse than going alongside a dock wall rather roughly in a gale of wind. Neither ship asked for assistance. The *Allegro* soon steamed off to Fawley as planned. The *Pacific Glory*'s crew put on lifejackets and swung out the lifeboats. But no general call to emergency stations was sounded and a quick external inspection suggested that the hull had been badly dented but not actually ruptured.

In fact, the curious sequence of events that cost the lives of fourteen crew members later that night had already begun. Chief Officer Kwong Fu Wai and the pumpman found the door into the cargo pump room at main deck level had been forced open in the collision. They could see nothing inside, but what they heard in the darkness disturbed them – the noise of oil pouring into the compartment from the starboard bulkhead. A check of the aftermost cargo tank on that side confirmed that the level was slowly falling. The crude oil was leaking into the cofferdam behind it and from there into the pump room.

In the engine room what appeared to be a more serious leak had developed, a mixture of oil and water draining down on the starboard side. The bilge pump was started and was obviously coping with the leak. But to keep it running, one of the two standby diesel generators also had to be running and no one was keen on that, least of all Captain Chang, because a strong smell of gas was beginning to pervade the engine room. His Chief Engineer took the point, but he thought there was no immediate danger. People were managing to carry on in spite of the fumes and without some source of auxiliary power the main engines would soon be flooded and useless.

The ship had two standby generators, installed one in front of the other on the port side of the engine room floor. Number 1 happened to be dismantled for repair at the time, so the engineers had no choice but to use Number 2, up against the forward bulkhead just beneath the ventilator shaft coming through from the pump room. It had been started up as soon as the main engines were shut down and it was kept going now, in spite of the Master's misgivings.

This caused no problem until about an hour after the collision, that is about 21.30, when the crude oil that had been quietly seeping into the pump room from the damaged cargo tank reached the level of the ventilator fan shaft. From then on a serious leakage of oil developed through the shaft, some of it spraying on to the generator pounding away fifteen feet below. Efforts to shield the generator from this oily spray were only partly successful.

The scene was now set for the disaster that occurred two hours after the ships collided. At 22.30 the Second Engineer noticed that the generator was beginning to accelerate. It would not respond to any local adjustment so he asked those manning the engine control room to shut it down from there. They tried to do this first by cutting out the electrical load, but it just made matters worse. Then they immediately shut off the fuel supply, but to their bewilderment the generator showed no sign of slowing down, on the contrary, it was still accelerating. The bearings screamed; the relief valves blew; black smoke began to pour out.

In retrospect, engineers were able to work out what had happened. The diesel generator had begun to function as a gas engine, fed by its own turbo-blower with the gas laden air of the engine room. The combustion cycle was operating with deadly efficiency and totally out of control. In seconds the generator started to vibrate on its mountings with such violence it could be felt throughout the ship. If anyone had stopped to watch, they would have seen the cylinder covers blown off, releasing incandescent gas to ignite the terrible explosion that wrecked the engine room.

Thirteen men died in the explosion or the fire that followed, four of them found in the engine room. Another six were picked out of the sea by the German vessel *Nordwelle*, although one of those died on the way to hospital. Twenty-three escaped in the port lifeboat.

The ship herself was still burning next day when she was beached near the Nab Tower at the eastern end of the Isle of Wight. The ruptured starboard side cargo tank was now leaking oil into the sea but some of this burned as it came out and a fleet of small craft was waiting to pump chemical dispersants on to the rest. The Dutch salvage firm of Smit eventually refloated her and towed her to dry dock in Rotterdam.

The subsequent Liberian board of investigation had no doubt about the navigational issues involved in the collision. It decided that the *Allegro* was never an overtaking ship under Rule 24 (now Rule 13) of the collision regulations, which states that:

> Every vessel coming up with another vessel from any direction more than 22½ degrees (2 points) abaft her beam, i.e. in such a position, with reference to the vessel which she is overtaking, that at night she would be unable to see either of that vessel's sidelights, shall be deemed to be an overtaking vessel.

It followed that in maritime law, the two tankers were simply crossing one another's course and the *Pacific Glory* was at fault for not keeping clear. But the board also blamed the *Allegro* for not taking more positive action under Rule 21 (now 17) once it was clear the *Pacific Glory* was not giving way. (It is this rule that has since been revised to give the stand-on vessel more discretion.)

Both crews were criticised for not keeping a proper lookout and for failing to plot the other vessel on their radar so as to establish that she was on a collision course. If the *Allegro*'s captain really had such difficulty in picking out the other tanker against a background of shore lights, the board said in its report, he had no reason to be blundering along at more than fifteen knots.

It is a comment that might have seemed equally apt a few months later, when there was a truly appalling series of wrecks further up Channel – the marine equivalent of a motorway pile-up, in which ships seemed to be ignoring not just a set of navigation lights but a whole array of pulsating green wreck marks, including a lightship's 250,000 candle-power lamp.

The pile-up began on 11 January with a collision that was bad enough in its own right, between the Panamanian tanker *Texaco Caribbean* and the Peruvian freighter *Paracas*. The tanker was bound down Channel in ballast, at night, when she met the dry cargo vessel coming up the English side of the Varne sandbank off Folkestone. According to a local fisherman who saw the Peruvian vessel pass, she seemed to be going the 'wrong' way just inside what was then no more than a recommended westbound shipping lane.

Anyway the two ships collided. The empty tanker exploded and sank, breaking up into at least four pieces, which then proved remarkably difficult to find as they were swept along the sea-bed by the strong tides. Trinity House had to mark the whole area with a pattern of green wreck buoys – the colour green being reserved in those days for marking wrecks.

The exact location of the *Texaco Caribbean*'s sunken bow was discovered a few days later by the German freighter *Brandenburg* – which also sank. Trinity House now had the equivalent of five wrecks blocking the main westbound shipping lane, some of them probably still on the move. It responded by replacing the spherical buoy on the north eastern corner of its diamond shaped pattern by a large lightship, flashing green every ten seconds. But it was not enough.

A few weeks later the Greek coaster *Niki* came down through the Strait, passed the frantically signalling light vessel, and ripped her bottom out on what was probably the sunken *Brandenburg*. Half the coaster's crew died abandoning ship in the cold water.

Trinity House was bitterly criticised by the *Niki*'s agents, who claimed that the previous wrecks were inadequately marked – which allegation the British lighthouse authority described as 'an outrageous slur upon men doing a difficult and dangerous job for the benefit and safety of seafarers of all nations'.

There was truth on both sides of this tragic argument. There can be no doubt that on a clear night, any seaman keeping half a lookout must have realised something was wrong when confronted by a cluster of flashing green buoys, let alone a lightship. Yet the hard fact is that many watch-keepers were unable to interpret the pattern of lights quickly enough. The first anyone knew of the *Niki*'s sinking, for instance, was a radio call from a nearby ship that had also sailed through the danger area, but with better luck. And ships went on sailing through it, even though Trinity House laid a further seven buoys and brought up a second light vessel.

It's not really surprising. Unless a passing ship happened to have heard one of the warnings broadcast by coastal radio stations, the man on the

Top The *Amoco Cadiz*. *Bottom* The salvage tug *Pacific*.

Top The 'driving school' for tanker captains, specially constructed near Grenoble. *Above* The hydraulic steering gear on a sister ship, identical to the system that failed. *Left* The chain which broke in the fairlead of the *Amoco Cadiz* when the first tow parted.

bridge would be taken totally by surprise when he saw the cluster of green lights beyond the lightship – or by day, a succession of green painted buoys. An isolated wreck would be marked by a single buoy, and shown on the chart. So all a navigator had to do was decide by the shape, or possibly the number of flashes, whether he should pass to port or starboard – and if in doubt he would just give it a wide berth. But off the Varne that winter he had to realise that he was dealing not just with one isolated wreck, nor even a group of separately marked wrecks, but a large danger area indicated by a coherent pattern he must interpret as a whole before his ship reached the first mark. Yet how else could Trinity House help him if they did not know exactly where every bit of wreckage was?

The new international system of buoyage introduced in European waters in 1977 abandoned the traditional colour green for wreck marks so as to use it for starboard hand channel buoys, and significantly adopted duplicated buoys to indicate a 'new danger'. With the new buoyage, a combination of the so-called lateral and cardinal systems, it should be possible to give more positive warning of any future Channel pile-up. But the immediate lesson drawn from the 1971 disasters was that they would never have happened had the Peruvian freighter followed the recommended route for through traffic along the French side of the Dover Strait.

Added to this specific point was a growing public suspicion, fed by the alarming press reports during that winter of 1970–71, that professional seafaring standards were in decline – if not in ships of the traditional maritime nations, then at least in those flying the dubious Liberian and Panamanian 'flags of convenience'. It was noted, for example, that neither of the Third Officers on watch when the two Liberian tankers collided held a Liberian certificate of competence, and the *Allegro*'s Greek Third Officer had no certificate of any kind.

Sailors no longer seemed to take elementary precautions like keeping a visual lookout, or slowing down when it was foggy, even though some of the vast ships they controlled were potentially floating bombs. It was at about this time that one heard the story, hopefully apocryphal, of the skipper who had trained his dog to stand in the bows and bark when he saw another ship coming. There was a new breed of ship's officer, it seemed, who preferred to stay inside his warm wheelhouse, occasionally glancing at a radar set which, far from avoiding collisions, sometimes helped to cause them. Why else should the experts talk about 'radar assisted collisions'?

Rightly or wrongly, casualties like the *Pacific/Allegro* collision and the *Texaco Caribbean* pile-up tended to confirm the lay public's worst fears. Members of Parliament in London sensed that it was time indignantly to ask

Government ministers what they intended to do about all this. Ministers turned to their officials with the same question, and were advised that controls must be tightened. If mariners could no longer be trusted to set their own safety standards on a voluntary, discretionary basis, standards would have to be imposed by the authorities ashore.

Most fundamental was the proposal that the recommended one-way shipping lanes, or traffic separation schemes, should be made compulsory. The British Government made the pioneering Dover Strait scheme mandatory for its own ships in 1972, and later that year IMCO produced a new set of collision regulations which would make all such schemes compulsory.

This move was fundamental in two senses. It was the first time anyone had dared to question the Master's traditional authority at sea (the same tradition that prompts many weekend yachtsmen to start bellowing like the ghost of Captain Bligh as soon as they step aboard their family cruiser). The professional mariner was used to accepting advice, or pilotage assistance. But in future this basic navigational decision would be taken for him, almost as if he were an airline pilot flying the airways under ground control.

The second point was that a new law of this kind would not be respected unless it could be enforced. And this meant that the authorities ashore must be able to police the traffic separation schemes. To some extent this could be done retrospectively, by penalising ships which collided while going the wrong way down a mandatory one-way shipping lane. But the British and French governments rejected this negative approach. They decided to experiment with radar surveillance that would at least give them a better understanding of the problem, and might then lead on to some degree of positive intervention by analogy with air traffic control.

The Dover Strait separation scheme therefore became just the first building block in a whole structure of surveillance and control. By the end of the decade there was not one shore radar in the Channel, but three, and six separation schemes, including one between Land's End and the Scillies, where the *Torrey Canyon* went aground, and another off Ushant, where the *Amoco Cadiz* came to grief.

The new Rule 10 of the 1972 collision regulations demanded that ships should generally either commit themselves wholeheartedly to a traffic scheme, and then stay in the appropriate one-way lane, or keep completely clear of it. If they were forced to cross the lanes – as ferries are in the Dover Strait – they should do so as nearly as possible at right angles. Fishing vessels and yachts should not impede ships using one of the schemes.

The regulations were also brought up to date by acknowledging the widespread use of radar and conceding that a deeply laden tanker should not be expected to run herself aground by giving way to a shallow draught vessel with plenty of room to manoeuvre.

Where there was a risk of collision, navigators were pointedly required to 'make proper use of radar equipment if fitted and operational, including long range scanning to obtain early warning of risk of collision, and radar plotting or equivalent systematic observation of detected objects'. A vessel constrained by her deep draught was invited to display three all-round red lights in a vertical line, in addition to her normal steaming lights.

To an extent, the maritime lawyers were merely trying to catch up with the reality of what was happening at sea in an age of supertankers. But that time lag was inevitable. A more serious criticism was that under the longwinded IMCO procedure, the regulations drafted and agreed in 1972 did not come into force until 1977. During the intervening five years a great many other things had happened to change the Channel's maritime environment, and they added up to a process for which someone coined a neat phrase – 'the erosion of command'.

Most professional mariners welcomed the change because they thought it would bring greater safety. Others reluctantly accepted that it was inevitable. A few doggedly resisted, arguing that the man on the bridge should never be dictated to by 'armchair sailors' ashore. In the early days of radar monitoring from St Margaret's Bay, the Coastguard's radioed warning that a 'rogue' vessel was heading the wrong way through the Dover Strait scheme would occasionally prompt a vigorous reply from the ship's Master concerned to the effect that he was well aware which direction he was going and if they didn't like it they knew what they could do with it.

The first radar had been installed in the small clifftop Coastguard station in 1971 after consultations with the French, Dutch and Belgian governments. Systematic monitoring started the next summer, shortly after the Dover scheme became mandatory for British ships, with VHF Channel 10 being used to broadcast warnings and navigational information. That year CROSS, a French equivalent of the British Coastguard monitoring system, set up its own radar station on Cap Gris Nez and was soon providing a complementary service from that side of the Strait.

To begin with, the radar screen in the darkened operations room at St Margaret's Bay merely provided an alarming confirmation that dozens of ships were ignoring the IMCO recommendations. But gradually the number of 'rogues' began to decline. From 1973 onwards the British Government began to prosecute a few each year, and issued official warnings to many more. The introduction of the new collision regulations in 1977, when routeing schemes became mandatory for ships of all the major flags,

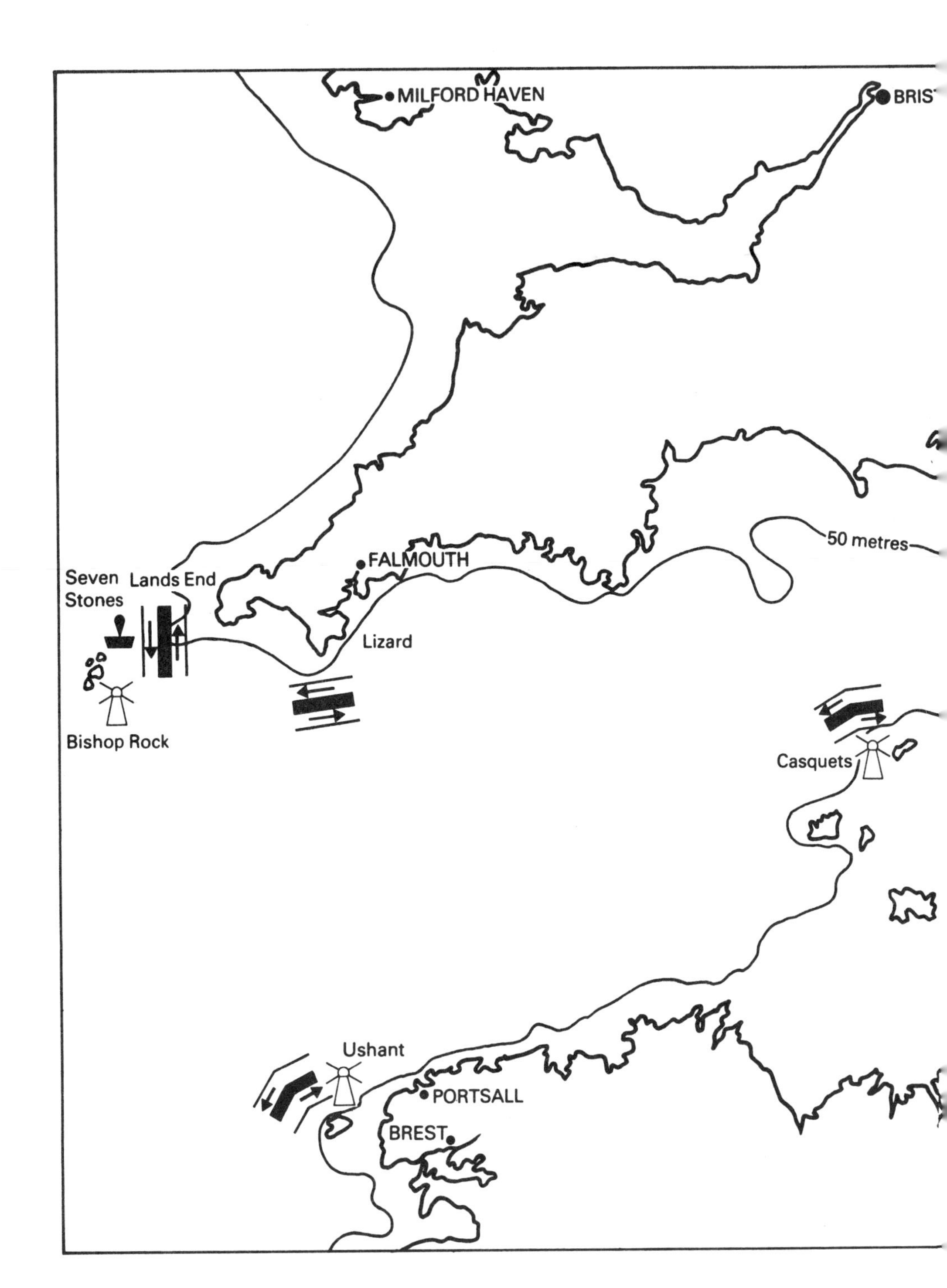
MILFORD HAVEN
BRIS
50 metres
FALMOUTH
Seven
Stones
Lands End
Lizard
Bishop Rock
Casquets
Ushant
PORTSALL
BREST

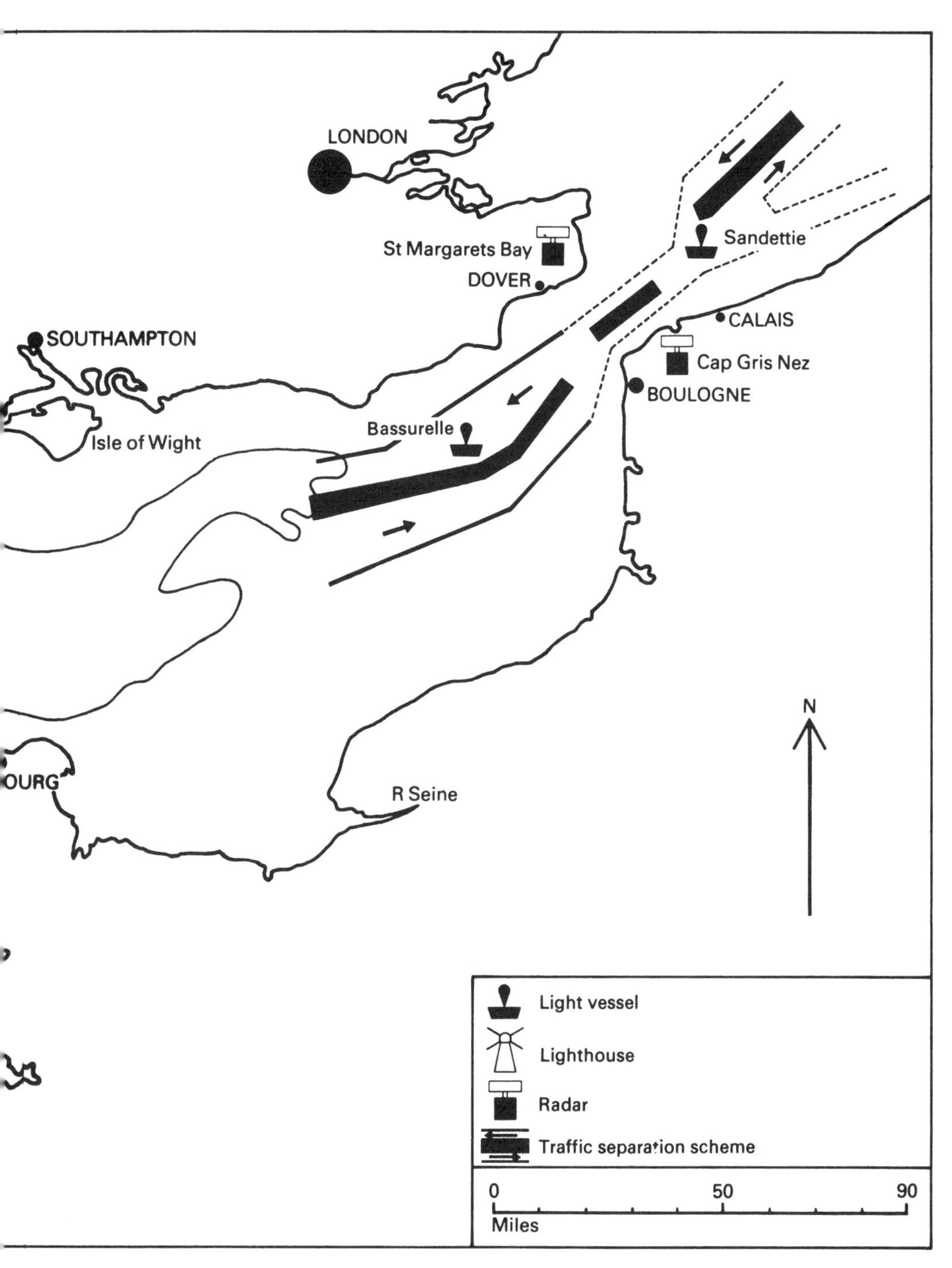

The separation schemes for shipping in the English Channel together with relevant navigational marks and the radars which monitor the traffic in the Straits of Dover.

brought the daily number of violations down to less than twenty, and many of these were local fishing boats.

Meanwhile the scheme itself had been extended and modified to provide for a special deep water route for loaded tankers past the Sandettie Bank. The British and French governments got together to co-ordinate their safety planning, and in particular to exercise a joint contingency plan – Mancheplan – to deal with any emergency from Dover to the Scillies. Two new 20 kw radars were installed on the British side at St Margaret's and Dungeness. The French studied the possibility of building one offshore on the Sandettie. On both sides of the Strait *ad hoc* monitoring arrangements developed into a smooth routine – plotting the rogues, organising light aircraft or naval patrol to identify them, broadcasting their course and speed to other vessels, together with weather forecasts and news of any other hazards to navigation such as a buoy out of place, a clumsy oil rig under tow, or a ship at anchor in an unlikely spot.

Two things hinted at the way the experiment would develop. Although both the British and the French referred to their monitoring as an information service, once routeing became compulsory the passive radar operator inevitably acquired something of a policeman's role, and at an international level. During the twelve months from July 1977, the British Coastguard observed more than 7,000 contraventions of IMCO's Rule 10, and reported more than 400 of them to the ships' respective governments. However, only a small number of these reports, notably to West Germany and Belgium, resulted in prosecutions. The other significant development was the Coastguard's acquisition of film recording equipment that photographed the radar display every minute. The film could then be speeded up to show, for example, the way in which the collision developed in February 1979 between the French passenger ferry *Saint Germain* and the Liberian bulk carrier *Artadi* off Calais. It even proved possible to confirm fears that a yacht which disappeared had actually been run down in the Dover Strait.

But of course this filmed record was a crude piece of technology by modern standards and at St Margaret's Bay the Coastguard were already looking ahead to new possibilities of analysis and control once they moved into their new, purpose-built operations room at Langdon Bay, just along the cliffs overlooking Dover Harbour.

5/The Neglected Province

JUTTING OUT INTO THE ATLANTIC like the prow of a ship, Brittany 'the land of the sea' in Breton, the ancient language of this originally Celtic land, has 688 miles of some of the most glorious coastline in Europe. Of that, 375 miles is the north coast along the English Channel and out to the Atlantic, stretching in rocks of subtle shades of pink, grey and mauve westward from Mont St Michel to Brest; a shore of sweeping bays giving way to a rugged grandeur made up of low-lying flooded coastal valleys, or *ria*, and protruding rocky headlands sheltering beaches of fine sand and huddled fishing ports. The *ria*, or *abers* as they are known locally, reach back inland towards the quiet, green chessboard of the Argoat, once covered by immense and impenetrable forests of oak and beech. Now the forests have given way to more gentle coppices, fields and lazy meadows where broom and heather stir in the wind, and deep lanes wander between high banks up to the central uplands where strange prehistoric monoliths stand looking out over the wild seascapes.

This beauty had produced natural riches for Brittany: the flora and fauna of the coastline (over 2,000 miles if every crenulation were taken into account) is one of the richest marine environments in the temperate world and the premier region of France for fish and seafood of all kinds. At least seven per cent of the nation's oysters came from one bay alone, and almost forty per cent of all the fish, including shellfish, in France came from the region. Twelve thousand fishermen, more than a third of those working anywhere in the country, fished from Brittany ports.

Along the northern coast for miles, seaweed was harvested for fertilisers, cosmetic products and animal feed, providing more than seventy-five per cent of the national algae production. In the fields behind, the farmers, famous for their pigs and artichokes, produced up to fifty per cent of France's agricultural produce in some crops. And the beauty brought the tourists, too: 4,500,000 in 1977, 723,000 of them from other countries, staying in hotels, guest houses, rented homes, cottages, camp sites, and in the boats bobbing in the many harbours.

As March of 1978 arrived, the two and a half million people of Brittany were desperately hoping for a good year. For in spite of the natural beauty of their homeland, the Bretons were not the happiest people in France. Like many remote country districts they had witnessed the drift from the land and the sea, and the migration of people to the more prosperous towns and cities of the nation. And they had had to seek ways of preventing the drift and bringing new work to the region. But no other area in France had had to undergo such an upheaval in so short a period of time. A total of 320,000 people, a seventh of the local population, had left agricultural jobs alone since 1954, and the exodus was still far from finished. Their country, like all those in Europe, had been going through troubled times with unemployment and inflation: in 1977 an average of 50,920 local people a month were looking for work, more than four and a half times the average of 11,164 across the ninety-six departments of France. Local leaders had a slogan: 'If you live in Brittany you are not in France' and spoke of the rest of the country as 'abroad', and it was more than simple Celtic pride. Despite the fact that major centres like Rennes and Brest were only a few hours on the train from the capital, the Bretons felt cut off, and that they were not receiving their fair share of help from the Gaullist-led coalition Government: and that in spite of the fact that Brittany was a strong rightist area. Two government ministers were local mayors, Yvon Bourges, the Defence Minister, at Dinard in the *département* of Ille-et-Vilaine; and Marc Becam, a Minister of State at the Ministry of the Interior, at Quimper, in Finisterre; supposedly giving the region a voice in the corridors of power. And yet not enough government aid seemed to have been forthcoming to an area where so many were out of work; where farms were rumoured to be closing at the rate of ten every day; and where there was a steady drain of young people moving east of a line through Cherbourg, Rennes and Nantes, because there was no work for them at home.

Though there had been some important achievements, especially in attracting new businesses such as electronics, telecommunications, chemicals and plastics, new industry did not seem to want to put down roots in Brittany, mainly because it was far from other centres and, thanks to government disinterest the Bretons insisted, communications were poor, with only one main rail line and only a plan for any of the new highways and autoroutes which spanned other areas of France.

And so the Bretons had learned to hold on to the natural advantages of the region: modernising their farming (which still represented twenty-five per cent of all employment); encouraging fishing and other marine ventures (nearly twenty per cent of all jobs); developing the quarrying of granite and kaolin; starting their own ferry line to ship their produce to international markets and to bring in tourists; developing the vital tourism which also provided good secondary employment; persuading government-backed research organisations to be set up in the region; rejuvenating some of their ports and regenerating the ship repair industry; concentrating on holding up employment in the Brest docks and the naval arsenal. In the twelve years up to 1974 they had succeeded in slowing the rate of migration and created 60,000 new industrial jobs, and 100,000 secondary jobs, through new investment and the strengthening of traditional forms of employment.

But since the oil crisis and the world recession the region had been in difficulty and the creation of a stronger regional economy and the slow drift of population had been held in an increasingly fragile equilibrium. As if to confirm their worst fears, the Government's latest economic plan concentrated upwards of eighty per cent of its new investment in four strong eastern areas of France. The last thing Brittany wanted was anything which would take away any of the advantages given to her by nature, or which might upset the delicate balance of the economy when the future was so uncertain. Jean Rouyer, president of the regional Chamber of Commerce and Industry summed it up in a sentence, 'For a region, nothing is more worrying than being stopped in mid-stream and not being able to see a horse to change to.'

Each year raised as many problems as the last for a land that had to live off nature and 1977 had been only about average. Rouyer himself had been emphasising the need to capitalise further on the rich waters surrounding Brittany, to utilise her ports more and to improve communications with the rest of the country to get goods to market; and everyone looked back to the classic good summer of 1975 when the *chiffre d'affaires*, the turnover, of almost every business had looked a little less depressing. So much depended on the weather and so far 1978 had not been good, with the winter seeming to want to hang on and wait for Easter. The fishermen, in particular, were in no mood to see their livelihood threatened. They had just been reminded by their union that some of them had still not received full compensation from the Government in Paris for their loss of income eleven years before, when 30,000 tonnes of oil from the *Torrey Canyon* had blackened Breton waters. Left wing candidates in the General Election looming for the middle of the month were already making great play of an ecological stand, knowing how important the sea and shore were to Brittany; and their platform, with unwitting prescience of what was to come, included promises to do something about the number of ships which kept coming to grief on the Finisterre rocks, and to protect the coastline from another *marée noire*, a black tide, like that of 1967. They knew that ecological candidates had made strong gains in the municipal elections in France in 1977.

Few people in Brittany would forget the *Torrey Canyon*. Within ten days of the wreck the winds and tides had begun to deliver 30,000 tonnes of oil on to the beautiful part of the Brittany shore known as the Pink Granite Coast of the Côtes du Nord. Marine life along the coast was devastated, fishing disrupted and thousands of seabirds from one of France's most important bird sanctuaries at Les Sept Iles, near Trégastel, died. And that had been just the start of a series of tanker accidents along the coast. Two years later on 19 August 1969, The French ship *Gironde* collided with another vessel and put 2,000 tonnes of oil into the sea. And on 24 January 1976, the brand new, and, mercifully, empty supertanker *Olympic Bravery* went on to the rocks north of Ushant. Only a few weeks later on 13 March, 1,250 tonnes of fuel oil from her bunkers leaked on to the shore. And only nine months later, north east of the Ile de Sein, near Brest, the German tanker *Böehlen* went down in a storm with 9,700 tonnes. About 1,000 tonnes went into the sea and after an attempt by the French authorities to pump more oil from her, she sank in 300 feet of water with most of her cargo still aboard. Of her crew of thirty-five, fifteen died and ten disappeared in the storm.

Though none of the other vessels had spread anything like the oil slick of the *Torrey Canyon*, with its economic and ecological consequences, the Bretons had made it abundantly clear to anyone who would listen that they feared that the sheer regularity of such accidents made it almost certain that one day something catastrophic would occur. In the first days of March, the walls of the students' quarters at the University of Western Brittany in Brest, and in the cafés and bars in the town where they met, were plastered with handbills and posters making the point graphically: most of them were headed with what became the symbol of all those who wanted something done about the *marée noire*, a stark picture of the dead body of an oil-smeared cormorant. And what angered them more than anything was not only feeling helpless to do anything about preventing disaster, but the knowledge that the next oil, when it came, would probably not even have been destined for delivery to France. The prow-like promontory of Finisterre was simply the last, hazardous, corner for tankers carrying oil to the whole of Europe and Scandinavia: 500 million tonnes a year, with eighteen vessels a day of more than 60,000 tonnes rounding Ushant. The most likely destination for tankers sailing the *Route du Cap*, as French navigators called it, was Rotterdam, a port which not only had a refinery capacity of 88 million tonnes itself, but was connected by pipeline to inland refineries in Belgium, Germany, Italy, other parts of Holland, and in France itself. And there were at least a dozen other refineries, particularly in

Britain and Scandinavia also served by tankers using the English Channel. It had not been forgotten that supertankers came round Ushant so heavily laden that they had to heave-to in Lyme Bay on the English coast and transfer some of their load to other vessels before they could proceed to Rotterdam.

But if the environment was an issue in the forthcoming General Election, it was not the only one occupying minds in Brittany that month. Most people were worried about what they saw as a much more fundamental threat to their way of life. The election promised to be one of the most crucial France had seen since the war, and the entire western world was watching fascinated from the sidelines: a heavy victory was being forecast for an uneasy coalition of the Left in which the Communists, led by Georges Marchais, might actually ride into power alongside Francois Mitterand's Socialists, and drive the Gaullist-led governing coalition into the wilderness.

Those who were concerned over legislation to provide better protection for the environment found themselves with a particularly agonising choice. They criticised as weak and haphazard the Government's recent initiatives and their response to tanker accidents and believed that the Left might be more willing to take a strong stand, both to protect the French coastline, and to spur action in slow-moving international treaty negotiations. But a vote for this mood would also be a vote for the rest of the Leftist platform, set out in the Common Programme, which had bound the parties loosely together since 1972, and which contained, for many, other less palatable propositions, including large-scale nationalisations of industrial concerns and banks.

France had a two-round simple majority system of voting for her General Election. And the dates chosen by the Government to go to the polls had a vital significance for what was to happen on the Brittany coastline. The first round 'primaries' open to all candidates for the 491 seats of the National Assembly, was fixed for 12 March, a Sunday, the traditional French polling day. As the voters went to the polls to choose their next Government, the *Amoco Cadiz* was already pushing north up the Atlantic coast of Africa. The second round of the vote was fixed for a week later, 19 March.

The morning of Monday, 13 March, the full results found the country in an incredible cliff-hanging situation. The Government had taken 46.5 per cent of the vote but the three major parties who had signed the Common Programme had scored a massive 45.1 per cent. The balance, if the vote should remain the same in the second round, now appeared to hang with the small parties outside the coalitions, who were suddenly clothed as

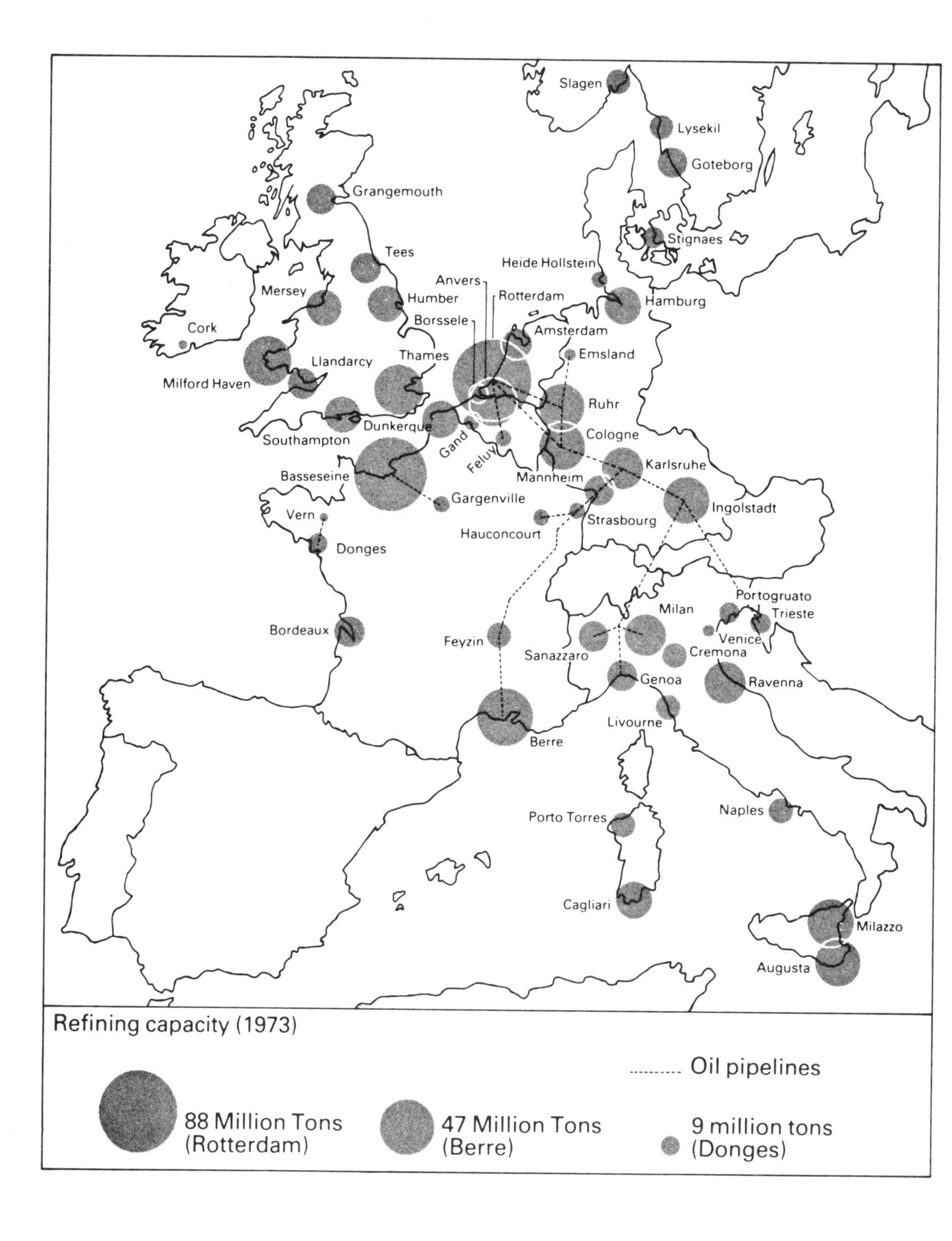
Slagen
Lysekil
Goteborg
Grangemouth
Stignaes
Tees
Heide Hollstein
Anvers
Mersey
Humber
Rotterdam
Hamburg
Cork
Borssele
Amsterdam
Thames
Emsland
Llandarcy
Milford Haven
Ruhr
Dunkerque
Cologne
Southampton
Gand
Feluy
Karlsruhe
Basseseine
Mannheim
Gargenville
Ingolstadt
Vern
Strasbourg
Hauconcourt
Donges
Portogruato
Milan
Trieste
Bordeaux
Venice
Feyzin
Sanazzaro
Cremona
Genoa
Ravenna
Livourne
Berre
Naples
Porto Torres
Cagliari
Milazzo
Augusta
Refining capacity (1973)
Oil pipelines
88 Million Tons (Rotterdam)
47 Million Tons (Berre)
9 million tons (Donges)

kingmakers: the extreme Left who had taken 3.3 per cent; and, significantly, the unregarded Brice Lalonde and the Ecologists, who had done well in the municipal elections of 1977 on a platform of protecting the environment, and who now held 2.1 per cent of the national vote.

Though the two Brittany majors, Yvon Bourges and Marc Becam, had been returned handsomely in their safe constituencies, this was not a week when any of the candidates in the Brittany area could afford to discount any voter asking for promises of legislation to protect the environment.

But in Brittany an extra and violent ingredient was added to the equation: the Breton separatists. Like fringe movements everywhere, they remained a shadowy organisation always ready to capitalise on a time of tension and confusion in the established order of things. Though no one knew who they all were, or even how strong they were, it was generally agreed that the more dangerous members were young, idealistic and middle class and had a forceful way of making their point. The Breton Revolutionary Army (ARB) and its so-called political wing, the Breton Liberation Front (FLB) had emerged from a number of nationalist and cultural groups and had been in violent action since 1966. In twelve years, in the name of an independent and socialist Brittany, they had carried out 206 bomb attacks, including blowing up the main Finisterre radio transmitter, Government and High Court buildings in Rennes, tax offices, barracks, the homes of deputies, and any other target which smacked to them of Paris 'oppression'. Only a month before, the first eight of fourteen FLB members to be arrested had appeared before the Court of State Security in Rennes and admitted thirteen different bombings in the province between 1975 and 1977. The most serious, in December 1975, caused over £1,000,000 worth of damage to the court-house in St Brieuc. Rounding up the network, who were eventually sentenced to up to eight years in prison, police had found seven rifles, revolvers, hundreds of cartridges, 150 pounds of dynamite, 400 detonators and 100 yards of fuse as well as timing mechanisms and masks. As if to prove that its power was undiminished, on the night the hearings opened other members of the ARB (which was said to have had three of its units trained in terrorism techniques by the Libyans and Algerians) blasted two tax offices, one in the Côtes du Nord and the other in Finisterre.

Though anarchistic, one thing the separatists agreed on was that Brittany

Left Europe's vast appetite for oil is largely fed by refineries on the continent's western seaboard. This diagram shows refinery capacity in 1973: in spite of North Sea oil the number of tankers passing through the Channel on their voyage from the Middle East has not diminished since that date.

would be better off on her own and they had recently been choosing the ecological platform to show how the Bretons had been failed by Paris. They also warned of the consequences of letting the shadow of the *marée noire* hang over a region so rich in natural resources that just one section of the coastline between Brest and Morlaix had been able to produce over £4 million worth of fish, crustaceans, oysters, scallops and harvestable seaweeds in the six months from March to August of 1977; and where tourists had been attracted to the region as a whole in sufficient numbers to inject almost £600 million into the economy that same summer. They had been looking for a way, as the election got under way, to make their point again. And, though they did not know it, as a gale blew up on the Finisterre coast in the hours after the first round election results, that opportunity was about to arrive.

Nominations closed for the second round at midnight on Wednesday 15 March. As the people of Brittany sat around their TV sets in the darkness of the evening, the wind and rain rattling on the window panes, the *Amoco Cadiz* was ploughing its way up across the Bay of Biscay in seas that could make even a 230,000 ton supertanker roll like a coaster. The TV news bulletins announced that the next day, President Giscard d'Estaing would make a final appeal to the electorate to 'use their intelligence' when they voted. In the meantime France went through the night and into the morning of Thursday 16 March with, constitutionally, no one on the bridge.

6 / Shipwreck

AT DAWN ON 16 MARCH the *Amoco Cadiz* was rolling north eastwards about thirty miles off Ushant, the southern 'gatepost' of the Channel, with an uncomfortable crossing of the notorious Bay of Biscay almost behind her. Captain Pasquale Bardari had had an uneasy night. At 1 am he turned into his bunk to try to get some rest before his ship closed land and entered the congested coastal shipping lanes. But he found real sleep impossible. For days past westerly gales had been blowing in that corner of the Atlantic, building up a heavy fifteen or twenty foot swell.

If the weather had been dead astern, the motion might have been easier. But on this course the swell was running in from the left, on the tanker's port quarter. On the bridge the helmsman had switched off the automatic steering and was rhythmically spinning the wheel by hand to check the twisting yaw as first the stern and then the distant bow was lifted bodily sideways.

At 4 am Second Mate Raimondo Salvezza noted in the bridge log: 'Wind WSW Force 7; sea very rough; heavy rolling; low visibility; radar on; sea on deck.'

Twice during the early hours Captain Bardari got up from his bunk to peer through his cabin windows, and called the watchkeeping officer to check weather and visibility. Shortly after 5.30 am he gave up hope of any more sleep and went on the bridge himself.

Up there things looked a bit more cheerful. The returning daylight revealed the enormous size of the seas, but they were easier to anticipate and it was also pretty clear that the wind would soon veer north westerly – the usual sign that a front is passing through, with the promise of brighter weather.

The British Meteorological Office's 0539 forecast for sea area Biscay promised 'Winds South West to West Force 7 to Severe Gale Force 9, locally Storm Force 10. Rain or showers. Visibility Moderate or Good'. But for sea area Plymouth, into which the *Amoco Cadiz* was moving, the forecast was: 'Winds North West Force 7 to Severe Gale Force 9, decreasing Force 5. Showers. Good Visibility.'

That meant that the rock-strewn coast of northern Brittany would be a lee shore – a fact that even the modern motorship sailor is instinctively conscious of. But Captain Bardari had no intention of staying on this side of the Channel. Mentally he was already charting a course of 037 degrees that would take him from the far end of the one-way shipping lane off Ushant across to the shelter of Lyme Bay, on the English coast. There he had orders to lighten ship by pumping some of his crude oil into another tanker before proceeding up Channel for Rotterdam, where he was to unload the rest.

But almost as soon as he reached the bridge there was a reminder that in his present position, the south westerly gale was still rising. A particularly large wave surged over the poop deck rail and carried away a stack of lubricating oil drums. To give the Chief Mate and his men a chance to get them safely under control, Captain Bardari ordered the helmsman to turn the *Amoco Cadiz* into the weather and eased back the engine revolutions, even though the ship was already making no more than about two thirds of her fifteen knot service speed. It took an hour and a half to collect the drums and stow them again securely, by which time the tanker was well to the west of her track. Course was altered five degrees to ease her back eastwards, ready to enter the northbound shipping lane at Ushant.

In fact, the chart used on the Liberian tanker's bridge that morning suggests that she did a good deal more than ease back to the eastwards. A line of five positions pencilled in between eight and nine o'clock, and all apparently plotted by the Third Mate Domenico Costagliola, taking a bearing from Ushant's Pointe de Créac'h lighthouse and a radar range, show her making almost a right angled turn across the end of the separation zone between the two shipping lanes before settling on to the correct northbound course. But under cross examination at the subsequent Liberian Government inquiry in London, the Third Mate made surprisingly little effort to justify his navigation. The ship never made such a turn, he said, so some of the points he plotted on the chart must have been wrong, either because he confused some other lighthouse with Pointe de Creach, or because he read the distance wrongly on the radar.

Captain Bardari later agreed that some of the points were 'a little out of line', but added another partial explanation. At about 8.30 am, he told the inquiry, he had run the ship off to the east for ten minutes to shield the crew while they recovered a fire hose and a couple of empty drums that had broken loose near the central deckhouse. At all events he then took over the navigational plot himself, and the *Amoco Cadiz* proceeded smoothly up the middle of the traffic lane until her progress was disturbed by another incident that was later to puzzle the board of inquiry.

At about 09.30, when he had been expecting to turn on to a course for Lyme Bay that would take him slightly offshore, Captain Bardari was forced to altar course in the other direction, to starboard, to make way for a small unidentified tanker heading south – a so-called 'rogue' vessel going the wrong way through the traffic system; and having altered course for one ship, he had to hold on until several more had passed.

So it was that at 09.45, when helmsman Fede suddenly called out that there was something wrong with the steering, the *Amoco Cadiz* was something like a mile closer to the French coast than her captain had earlier intended – that is about eight and a half miles north of the island of Ushant and roughly fifteen miles from the mainland to the east. That lost mile was later to prove fatal.

The ship had just begun one of her slow rhythmic lurches to port, and the helmsman was turning the wheel in the opposite direction to check the incipient turn. But with twenty degrees of starboard wheel applied, he realised that not only was the giant tanker failing to respond; the rudder indicator showed that the rudder was hard over the wrong way, to port. In short, she was out of control.

Captain Bardari responded quickly and decisively. In rapid succession he telephoned the engine room to tell them they had a rudder problem, reached for the telegraph to ring 'Stop Engines', ordered two black balls hoisted on the flying bridge to indicate that his ship was 'Not under command', and put out a VHF radio call, direct from the bridge on Channel 16, to warn other shipping. His broadcast named the *Amoco Cadiz*, gave her approximate position and warned other vessels to keep clear because she had steering problems.

The warnings were certainly called for. A loaded 200,000 ton supertanker does not simply stop when you stop the engines. Even with them full astern, she ploughs on for two and a half miles before losing way. But in this case, with the rudder initially jammed over to port, she veered off to the north in a great one and a half mile arc, right into the path of ships heading across to join the southbound traffic lane. If another ship's navigator had his binoculars conscientiously trained on the Liberian tanker's signal halyards, he would of course realise that something was wrong. But it would have been an alarming sight, to say the least, to come upon her suddenly emerging from one of that morning's periodic rain squalls, sheering wildly off to port, when she would normally have been expected to give way to crossing traffic by shifting to starboard.

It was twenty minutes before the *Amoco Cadiz* finally stopped, about a mile and a half further north. She was slewed right round facing south east, towards the shore, rolling broadside to the long Atlantic swell. Captain Bardari left the Second Mate in charge on the bridge and headed aft to see for himself what was wrong in the steering compartment. He was accompanied by the Chief Mate and by Lesley Maynard, a Marine Safety Services specialist who had joined the ship in Las Palmas by helicopter, together with the relief Chief Engineer Antonio Assante.

Assante was not due to take over from the existing Chief Engineer Salvatore Melito until the ship reached Rotterdam. But that morning, as part of the handover procedure, both men had been wandering through the noisy innards of the tanker on a joint tour of inspection. They had just checked the evaporators in the refrigeration plant and Assante was leading the way back up to Melito's cabin. But when he got there, his companion was no longer following. Assante waited a moment or two and then, with an engineer's habit, glanced at the ship's engine revolutions indicator. It was reading zero.

Running down towards the engine room he met a fireman who told him the trouble was in the steering compartment, so he went straight there. He found the floor covered in oil – hydraulic fluid from a ruptured junction box on which the Chief Engineer and his First Assistant, Michele Calise, were already working to shut down the valves with a handwheel and an eight-inch wrench. It was the start of an hour and a half's struggle by the entire engineering team – first to repair the steering mechanism and then later, as the great barn door of a rudder began to take charge of the system it was supposed obediently to serve, to prevent the gear being smashed to pieces. The slippery, heaving floor did not help; nor the dreadful noise of the rudder slamming back and forth in this sounding box of a compartment as the ship lost way.

The steering gear on the *Amoco Cadiz* consisted of twin electric pumps, driving two pairs of hydraulic rams attached to each end of a balanced tiller bar on top of the rudder's vertical stock. When the hydraulic rams turned the tiller in response to the wheel, the submerged rudder blade turned in the same direction. But on the mammoth scale of this ship, the term 'blade' was scarcely appropriate; it consisted of a flat steel box more than forty feet deep and nearly thirty feet broad, which when struck by a heavy wave was capable of transmitting enormous pressures back through the hydraulic system.

The failure which eventually led to disaster began in quite a small way. Five bolts (or, strictly speaking, studs) holding a pipe from the port side pump on to the hydraulic distribution block broke off, allowing the pressurised fluid to force back the flanged end of the pipe and escape in a powerful jet. The engineers' first action apart from dodging the oil was to switch power from the port side pump to the starboard one and then close down the valves isolating the damaged side of the distribution block. That stopped the flow of oil and in principle they still had a system capable of driving the rudder.

But they also had a lot of air in the system, sucked in through the hole where the oil escaped. So their next job was to bleed the system, rather as one bleeds the brakes on a car, by opening valves on the hydraulic rams and letting the air out as the cylinders refilled with oil from the header tank. Simple enough in the repair manual, but not so with the rudder slamming backwards and forwards, shunting an unpredictable mixture of air and fluid violently through the network of piping, valves and cylinders.

In fact the engineers struggled on for an hour, opening and shutting valves, sloshing buckets of oil into the header tank, but they were getting nowhere. Then another pipe failed, smaller but more critical. It connected a safety valve with the top of the distribution block, and this time there was no simple way of isolating it. From this moment the engine room team gave up hope of restoring the system to operation and concentrated instead on merely trying to fix the rudder in one position or another, controlling the fearful spasms with which the machinery was literally destroying itself. The second burst pipe had released what was left of the cushioning in the system and the rudder was now slamming right up against its stops, metal to metal. Someone fetched a chain tackle capable of holding a five ton load, but the rudder head snapped it like a piece of string and sent the man off to his cabin with a nasty gash on his head. First one tiller linkage gave, then the other. The great rudder blade that had done all this damage was now totally free to swing whichever way the waves took it, and totally useless to Captain Bardari on his bridge. Chief Engineer Melito went up to tell him the news at 11.20.

The captain had anticipated his chief's formal report ten minutes earlier by calling up Brest coastal radio station and asking where the nearest tugs might be. The reply was reassuring. The German tug *Pacific* was already at sea in the area and at 11.20, when the steering gear's breakdown was confirmed, Bardari went back into the radio room and asked Brest to summon her assistance. By 11.30, while he was still busy with a series of radio telephone calls to the Amoco offices in Genoa and Chicago, the bridge informed him they were in direct contact with the *Pacific*, which turned out to be only about fifteen miles – that is about an hour – away, and heading for their position.

There was relief all round. While the engineers were still trying to repair the steering gear, its failure had seemed like no more than a temporary interruption of the voyage. But with the ship permanently crippled, everyone on deck was sharply aware that the notoriously dangerous coast of Brittany was already visible between the rain squalls. Periodic positions plotted on the chart showed that the *Amoco Cadiz* was steadily drifting towards the shore, the nearest point of which was only 7½ miles away. The tanker's bow was also pointed inshore, heading about south south east. This was disturbing for important practical reasons. It meant, for example, that the engines could not be started ahead, even simply to stabilise the rudder while the crew were trying to secure it, without driving the tanker closer to the rocks. As for going astern, a ship's propulsion system is much less efficient in that direction and Captain Bardari was fearful anyway that it would cause more damage.

In those circumstances, the fortuitous presence of what sounded like a powerful tug – 10,000 horse power, apparently – must have been deeply reassuring. Yet when the lawyers got to work at the subsequent inquiry in London, cross-examining Bardari, Maynard, and the other members of the crew, they would suggest that it was a false reassurance, verging on complacency.

From this point in the story, it is impossible to divorce any reconstruction of the events of 16 March from the accounts that were later given to that inquiry in the lecture hall of the Royal Institution of Naval Architects – because there were two quite different accounts that began to diverge from the moment VHF radio contact was established between the two ships. The first is an amalgam of the Italian captain's evidence and the handwritten log kept by British safety expert Lesley Maynard that was recovered from the wreck with some – but not all – of the bridge records on a few sheets of oil stained paper. The second was the account given by the German tug skipper, Captain Hartmut Weinert, which flatly contradicted the tanker men's story on certain crucial points.

Lesley Maynard started writing his log at about the time the tug's assistance was called for. Most of it was dictated, though not word for word, by Captain Bardari. But Maynard was directly involved in some of the situations. His advice was evidently valued and he was often the man – 'speaking very fluent English' – with whom the tug skipper and his radio operator found themselves dealing when they called up the tanker. Keeping a careful log, with times and positions, is a professional seaman's habit. But by now both men on the crippled vessel's bridge were aware that the Italian officer would later be called on to justify his actions and that a great deal of money, if nothing more, might depend on them.

Their sea-going background was different. Bardari was a graduate of the nautical college in Pisa; he first went to sea as a cadet aboard a passenger liner, the *Gulis Cesari*; later he served on a succession of tankers, many of their names carrying the prefix *Amoco*. Maynard joined the training ship *Arethusa* at the age of thirteen and then worked his way up through the Royal Navy to become a specialist mine counter measures and clearance diving officer before leaving to join the Marine Safety Services consultancy. Maynard was on board for the last part of this trip to give a series of lectures on safety and to check out the crews' drills on lifeboats, fire fighting, breathing apparatus and so on. As the crisis developed off Ushant, he stayed with the captain on the bridge, trying to help without interfering, and he stuck with him that night when the ship was beginning to break up and the rest of the crew had been lifted off by helicopter.

At the board of inquiry, too, he supported the Italian captain, generally corroborating his evidence. Their joint account sounded totally plausible until Captain Weinert appeared to give an equally plausible, but quite different, version. As the conflict developed under cross examination by the Liberian Government's huge American lawyer, Dr Frank L. Wiswall, each of the three seemed to acquire a touch of national caricature. Bardari, for all that his master's licence was at stake and that he still faced the possibility of criminal proceedings in the French courts, seemed to find a certain sad humour in the situation. Maynard was briskly naval, snapping his answers to attention, and remarking at one point that he managed to get the gist of what the Italian officers said to one another in their own language from the way they waved their hands about – 'because they gesticulate when they are talking on the telephone as well'. Then Weinhart turned up – stocky, blond, blue-eyed – looking and sounding every bit as Germanic as the slim, smiling, dark-haired Bardari was Italian.

The first direct entry in Maynard's log, as opposed to a summary of previous events, marks the appearance of the *Pacific* alongside:

12.20 TUG ARRIVES AND PREPARES TOW

Everyone on the tanker's bridge was delighted to see her, circling under their stern and sizing up the job ahead. Both Maynard and Bardari, by their own account, thought she looked powerful enough to pull them clear, or at least to turn the tanker round so that she could use her own engines.

Indeed by any standard the *Pacific* was a hefty, purposeful looking vessel. Her high bow swept down to a low working deck, occasionally awash now as she rolled across the seas. Her towing gear was laid out ready – 1,400 metres of 6½ inch wire coiled on the winch drum, a heavy wire towing pennant with a soft Flemish eye spliced into the end, a double rope 'spring' to absorb shock loads, and two short lengths of special towing chain with a diameter of 65 mm. Her twin screws, thrusting through what are known as Kort nozzles, were reckoned to give her a 'bollard pull' of sixty tons.

The *Pacific* was operated by the Hamburg firm of Bugsier, either on ocean towage jobs, like moving oil rigs and taking laid-up ships to the breakers yard, or on salvage. In the latter case, her twenty man crew all received a percentage of the salvage award.

That morning they had been lying in the harbour in Brest, keeping their usual watch on weather and radio, when Captain Weinert received orders to proceed immediately to the Dover Strait. There he was to assist the tug *Wotan* to manoeuvre an oil drilling platform through the narrows and down to the Bay of Biscay. He set off at his best speed of 14½ knots and was well through the Chenal du Four inside Ushant at 11.25 when his radio operator took down a VHF radio message: '*Amoco Cadiz* to all ship. Our position eight miles north of Ushant. We have completely fallout of gear. Please keep clear.' Weinert told him to make contact and offer salvage under Lloyd's open form agreement, eased the tug's engines to Half Ahead as she turned into the westerly gale, and then rang Full Ahead for the position indicated by the tanker.

The dispute between the two captains began while the *Pacific* was still smashing her way through the head seas on her way to the casualty. The first response to the tug's offer of Lloyd's open form was a request for her simply to stand by. Half an hour later, according to Captain Weinert, he repeated the offer and 'their reply was that they were telling us to get in touch with Amocoship Chicago'.

The point about Lloyd's open form is that it is a standard agreement that provides for subsequent arbitration of the salvage award in London, under the auspices of Lloyds, and with the right of appeal. It is well known to seafarers all over the world and Bugsier always used it except for German and Scandinavian vessels, when its tugs would offer arbitration by the German Maritime Arbitration Court in Hamburg. Its advantage is that it can be agreed instantly, if necessary just by a few shouted words between the bridges of the two ships. There should be no need for haggling. The ship in trouble cannot be held to ransom and the rescuer knows that once he has his agreement his efforts will be rewarded according to the value of the property he saved, the danger it was in at the time and the risks he took to save it.

But of course the open form is a *salvage* agreement, and in spite of the appalling weather, the menacing coastline and his broken steering, Captain Bardari was determined not to admit that he needed salvaging. For one thing he still had full power in his engines, if only he could turn his ship offshore into the wind. In his account of the radio exchanges with the *Pacific*, endorsed by Maynard, she quite clearly accepted that this was a straightforward towing job. The German tug would merely assist the Liberian tanker to reach Lyme Bay, maintaining her heading while she used her own engines.

Many factors may have been in Bardari's mind at the time, including whatever conversations he had with Amoco and his own feelings of professional pride. But important among them must surely have been the thought that a towage contract would be measured in thousands of pounds or dollars, while a salvage award might run into millions. Whether he was right about that is a different matter. The terms of the towage contract were never discussed, he admitted to the board of inquiry, but he was in no doubt that he had a contract and 'of course it cost less'.

According to Maynard, the crucial conversation took place between 11.30 and 12.20: 'There was a radio call to us from the tug *Pacific* and the tug asked us for our geographic position, which was passed to him. He also said "I take it this is Lloyd's open form" and our captain replied "No, I want a tow mileage rate to Lyme Bay". . . . The reply to that from the tug was "OK".'

There was of course plenty of room for misunderstanding in these exchanges, particularly since some of them were being conducted through intermediaries at both ends. But Weinert denied that he was even asked for a mileage contract, although he did admit through his interpreter that by the time he reached the *Amoco Cadiz* his offer of salvage had still not been accepted: 'I took stock of the situation. I informed the vessel that they were in my view in a very bad position and in need of a tug, and thereupon the captain replied "Yes, we are indeed, but it would be best first of all to get a line across and then we can talk about the type of contract".'

So for the moment the talking stopped and both crews got down to the difficult, sometimes dangerous, job of connecting a tow line.

Connecting up the heavy towing gear needed for this job demanded a great deal of skilled co-operation between the two crews. Both ships were rolling and heaving in the big swell and now that the tug was close alongside the tanker's starboard bow, the extent of the vertical movement was alarmingly obvious. Occasionally one or other vessel would lift a lump of green water over the rail, to surge across the deck threatening to sweep the men off their feet. Much of the gear was too heavy to work by hand but dangerous to be near when it was under the tension of a powered winch. One flick as a cable snatched taut would be enough to break a man's leg.

LLOYD'S

NOTES.

1. Insert name of person signing on behalf of owners of property to be salved. The Master should sign wherever possible.

2. The Contractor's name should always be inserted in line 3 and whenever the Agreement is signed by the Master of the Salving vessel or other person on behalf of the Contractor the name of the Master or other person must also be inserted in. line 3 before the words "for and on behalf of". The words "for and on behalf of" should be deleted where a Contractor signs personally.

STANDARD FORM OF

SALVAGE AGREEMENT

(APPROVED AND PUBLISHED BY THE COMMITTEE OF LLOYD'S)

NO CURE—NO PAY

On board the
Dated /19

† See Note 1 above

IT IS HEREBY AGREED between Captain† for and on behalf of the Owners of the " " her cargo and freight and * for and on behalf of (hereinafter called "the Contractor"):—

* See Note 2 above

1. The Contractor agrees to use his best endeavours to salve the and/or her cargo and take them into or other place to be hereafter agreed. The services shall be rendered and accepted as salvage services upon the principle of "no cure—no pay". In case of arbitration being claimed the Contractor's remuneration in the event of success shall be fixed by arbitration in London in the manner hereinafter prescribed: and any difference arising out of this Agreement or the operations thereunder shall be referred to arbitration in the same way. In the event of the services referred to in this Agreement or any part of such services having been already rendered at the date of this Agreement by the Contractor to the said vessel and/or her cargo it is agreed that the provisions of this Agreement shall apply to such services.

2. The Contractor may make reasonable use of the vessel's gear anchors chains and other appurtenances during and for the purpose of the operations free of expense but shall not unnecessarily damage abandon or sacrifice the same or any other of the property the subject of this Agreement.

3. The Master or other person signing this Agreement on behalf of the property to be salved is not authorised to make or give and the Contractor shall not demand or take any payment draft or order for or on account of the remuneration.

PROVISIONS AS TO SECURITY

4. The Contractor shall immediately after the termination of the services or sooner notify the Committee of Lloyd's of the amount for which he requires security (inclusive of costs, expenses and interest) to be given. Unless otherwise agreed by the parties such security shall be given to the Committee of Lloyd's, and security so given shall be in a form approved by the Committee and shall be given by persons firms or corporations resident in the United Kingdom either satisfactory to the Committee of Lloyd's or agreed by the Contractor. The Committee of Lloyd's shall not be responsible for the sufficiency (whether in amount or otherwise) of any security which shall be given nor for the default or insolvency of any person firm or corporation giving the same.

5. Pending the completion of the security as aforesaid, the Contractor shall have a maritime lien on the property salved for his remuneration. The salved property shall not without the consent in writing of the Contractor be removed from the place of safety to which the property is taken by the Contractor on the completion of the salvage services until security has been given as aforesaid. The Contractor agrees not to arrest or detain the property salved unless the security be not given within 14 days (exclusive of Saturdays and Sundays or other days observed as general holidays at Lloyd's) of the termination of the services (the Committee of Lloyd's not being responsible for the failure of the parties concerned to provide the required security within the said 14 days) or the Contractor has reason to believe that the removal of the property salved is contemplated contrary to the above agreement. In the event of security not being provided as aforesaid or in the event of any attempt being made to remove the property

15.1.08
3.12.24
13.10.26
12.4.50
10.6.53
20.12.67
23.2.72 Ag. 13-D.C.

To fire the first line across, the tug men used what is known as a Konsberg gun – safer than a rocket where tankers are involved. The team dodging the waves on the forecastle of the *Amoco Cadiz* had to grab this line as it fell and pass it through the enclosed Panama fairlead right in the bow. Attached to it was an ordinary heaving line of the kind you see thrown ashore when a ship comes alongside a quay, followed by a five inch circumference messenger line. Then they had to use the steam-powered anchor winch to haul in a wire messenger, attached to a heavier wire pennant and a short length of immensely heavy towing chain – each link of it more than a foot long. The purpose of the chain was to take the special twisting strain where the tow passed through the bow fairlead. Once positioned according to the tug's instructions (radioed to the tanker's bridge and passed on by walkie-talkie) the pennant wire was made fast with ten turns round the bitts right forward and another six on the starboard side. Chief Mate Strano informed the bridge that the tow was secured and this fact was duly recorded in the Maynard log:

13.31 LINES ALL FAST

As soon as he was informed of this Captain Weinert began easing the *Pacific* away almost at right angles to the tanker's heading, that is straight into the weather, paying out the towing wire behind him. At 800 metres he stopped, but seeing that the wire was still whipping up and down on the deck, he let out another 200 metres. Shortly after two o'clock, nearly two hours after the tug's arrival and with the tanker now only 5.7 miles from the shore, his efforts began to show results visible to the men anxiously watching from the bridge of the *Amoco Cadiz*:

14.05 TUG PULLING. SHIP TURNING SLOW

It was a moment of great satisfaction for both crews. For one it signalled escape from potential shipwreck; for the other the prospect of a lucrative salvage award. But the euphoria did not last long. The bow had only turned through a few degrees to starboard when it began to swing back, turned again and then reverted more or less to its original heading – or so it seemed from the tanker's bridge:

14.35 IT WAS OBSERVED THAT THE TUG HAD STOPPED PULLING. TUG REQUESTED LLOYDS OPEN AGREEMENT. AGREEMENT REFUSED. TUG TOLD TO CONTACT CHICAGO. TUG THREATENED TO RELEASE TOW.

The Lloyds Open Form salvage agreement. This copy bears the company stamp of the Dutch tug company Smit, which was later brought in to attempt to salve the wreck of the *Amoco Cadiz*.

Now the argument about salvage terms had broken out all over again, and with a rising anger on both sides.

Weinert claimed later at the inquiry that once he realised the *Pacific* was not powerful enough to turn the deeply laden tanker into the wind he made repeated radio calls to find out which way the Liberian vessel's rudder was turned (not knowing that it was simply slamming about) but got no reply. He had not stopped pulling, even if it looked that way. The *Pacific* had eighty per cent power on all the time. Any more would have risked breaking the line and at that level he was managing to slow the tanker's drift, holding her until a second, more powerful Bugsier tug, the *Simson*, arrived to help.

On board the *Amoco Cadiz* they did not know about the *Simson*, which had actually been diverted at 11.58 that morning when she was scudding up Channel with the gale behind her. Captain Bardari, or so he told the inquiry, was convinced that now Weinert had his line aboard he was putting on the pressure to change his towing contract to a salvage agreement. Bardari, too, claimed that many of his radio calls went unanswered by the tug, but that when he did get through, at 14.35, he immediately gave way to the threat. 'You begin towing; I will inform Chicago' – by which he meant that he accepted Lloyd's open form and would inform his owners accordingly.

If a threat was made it was a completely empty one, because the German tug skipper had nothing to gain and a great deal to lose by casting off his tow. Certainly he denied making it, but then all the exchanges during this extraordinary argument – with 223,000 tons of black oil poised off the Breton beaches – are confused by not knowing for certain who was speaking to whom. None of the characters in this drama ever saw one another, except perhaps as distant figures standing on a bridge wing. At one point the German captain startled the subsequent inquiry by claiming that it was not until 15.15 that he spoke directly to Captain Bardari for the first time:

'Captain, you are in a very bad position,' he recalled saying in his guttural English. 'You have a very big ship. The weather condition is the same, very bad, and we must have Lloyd's open form. You accept it. Please you accept it.' To which the Italian captain allegedly replied: 'No, no, no.'

But on the tanker's bridge, the conversation apparently sounded quite different. Bardari was scuttling back and forth, trying to make radio contact with his managers in Chicago and repeating his demands for the tug to 'start pulling'. At about 15.45, apparently, he at last got through to Chicago and spoke to Captain Phillips, the Manager, Marine Operations and ten minutes later he was able to inform the tug that head office had approved his change of contract.

That, one might think, should have settled the matter. But by now, it seems, neither captain was prepared to trust the other's word. Weinert demanded an exchange of telegrams through Brest. Bardari complied, as recorded in the radio log: 'Captain German tug *Pacific*. We agree Lloyds open form salvage agreement. No cure no pay.'

It was now four o'clock in the afternoon. An hour and a half had been spent in apparently futile argument and the weather was still deteriorating. From the tug's point of view the time had not been wasted because even if she could not turn the big vessel's bow to seaward, she had continued to pull in that direction and therefore to slow the shoreward drift.

But the failure of communication between the two captains may nevertheless have been vital because Bardari seems to have believed the tug could do better if she tried. He was still relying on the *Pacific* to pull him clear eventually and in particular he had not yet used his own engines. A quarter of an hour later all that changed, when the tow line parted at what might have been thought its strongest point, the special chain at the tanker's bow:

16.15 TOW PARTED. TUG WAS NOT PULLING AT TIME. SHIP THEN ENDEAVOURING TO MANOEUVRE ASTERN TO KEEP CLEAR OF FRENCH COAST ROCKS RANGE 6.4 N.M.

Even above the noise of the gale Captain Bardari recalled hearing a 'wrenching' sound as the heavy chain was torn in two. Nobody could be sure why it had failed. Bardari himself suggested that perhaps it became twisted with the roll of the ship and then parted when the tug began to pull more strongly. Maynard had a vivid memory of the two ships being thrown apart by two huge waves: 'The tug went racing off down one, we fell to port off the other, and the line parted . . . we were all rather shocked and surprised and dismayed that the tow parted, and there was one of those long thinking silences. A few swear words were uttered forth, I think. Then we called the tug and said "The line has parted" and he said words to the effect of "Yes, agreed",'

As soon as Weinert heard where the break was he felt sure he knew the cause of it. The edge of the Panama fairlead was wrongly shaped. He reckoned it formed too sharp an angle, round which the chain was stressed. In other words it did not provide a fair lead at all.

But none of these rival theories mattered much at the time. The tug immediately began winching in to prepare a new tow. Aboard the tanker, Bardari reached for the engine telegraph. He rang Slow astern, Half astern, and then asked the Chief Engineer for all available power. The rocky coastline was clearly visible downwind, and nearer. He was going to attempt to back the *Amoco Cadiz* off even if it tore the rudder right out of her.

Unfortunately, the effect was less than dramatic. The rudder held, but spoiled the efficiency of the single propeller. The engines did slow the ship's eastward drift, which was obviously useful, but they could not overcome it. They also had the effect of turning her stern first into the weather so that the bow's new heading – about 130 degrees, or straight onshore – was even further from the direction she needed to make her escape.

The tug's first estimate of how long it would take to rig a new tow, according to the tanker men, was about an hour. But it took nearly that long simply to winch in the damaged gear, and each time the *Amoco Cadiz* radioed to ask how much longer, she was told 'another half hour'. The only comfort to be found on the tanker's bridge was from the chart, on which her position was plotted with increasingly anxious frequency. The direction of drift had definitely changed and she was now moving north eastwards, almost parallel to the coast. Then the tug radioed two encouraging bits of news:

> 18.00 PACFIC INFORMS CADIZ HE IS GOING TO ATTEMPT TO TOW US ASTERN AND HOLD US UNTIL 2ND TUG SAXONE (this was actually the Simson) ARRIVES APPROX. MIDNIGHT.

The *Simson* was a really big tug apparently, much more powerful than the *Pacific*. If the *Amoco Cadiz* could only scrape past the Roches de Portsall she could probably hold clear until midnight. It meant that even in these desperate circumstances, when Captain Bardari telephoned his office in Chicago at 18.35 he was able to offer a slim hope of saving his ship. But if it came to the worst, he told them, he would try dropping the anchors and put out an SOS general distress call.

Bardari's most immediate worry was the *Pacific*'s proposal to make this second towing attempt from the tanker's stern. At first he refused, trying to insist that it should be from the bow again, but the German tug took no notice and he was no longer in any position to bargain.

It was an extraordinarily inappropriate moment for the argument between the two captains to be rekindled. Yet it was natural enough. The Italian still evidently believed that the *Pacific* was capable of pulling his ship's head round so that he could use his own engines properly. Indeed it was already beginning to come round of its own accord and at 18.43 he tried easing the revolutions to 'Slow ahead' in the hope of encouraging the movement. The German had at least four factors in mind, as he explained at the London inquiry. He still did not know the position of the tanker's rudder. He was worried about the sharpness of the bow fairlead. He knew positively that his tug was not powerful enough to turn the deeply laden vessel into the wind the first time (although the *Simson* would be able to do it if he could meanwhile help the tanker clear) and he was acutely aware of the narrowing gap between the crippled ship and the shore.

As the tug finally manoeuvred close under the giant tanker's stern to connect the new tow, at 19.00, the nearest rocks on the Breton coast were only five miles away, marked by the heavily breaking seas. The submerged Roches de Portsall – just as dangerous to the *Amoco Cadiz* in her heavily laden condition – were less than four miles distant and the gale had now veered north westerly, driving her straight on to the lee shore.

The last thing Captain Bardari wanted was to stop the engines that had so far kept him clear, but if he was to accept the *Pacific*'s help he had to gamble on doing just that:

19.01 TUG REQUIRES STOP ENGINE – ENGINE STOPPED.

Captain Weinert waited upwind until the tanker was once again dead in the water, then eased his tug as close as he dare before firing his line across the stern where he could see the crew waiting to grab it. Bardari, watching impatiently from the wing of his bridge, knew that all this took time, and careful judgement on the tug skipper's part, but he could not resist reminding Weinert that time was fast running out:

19.15 TUG ADVISED WE ARE DRIFTING TOWARDS THE SHORE.

The first line slipped into the sea. Chief Mate Strano and his men managed to grab the second at 19.25 and started to get the messenger aboard but there was not enough slack in the heavier line and three minutes later it was torn out of their hands to slide back over the side. Now, to the tanker men's surprise, the tug repositioned herself on the port side of the stern. Weinert said later he could see the crewmen sheltering there. But with her engines stopped the *Amoco Cadiz* was slowly swinging broadside to the weather, so the Konsberg gun was now firing into the teeth of the gale. Time after time the line fell short. Captain Bardari decided he could wait no longer. At 20.04 he ordered Second Mate Vaudo on the bow to let go the port anchor. In the growing darkness the occulting light of the Roches de Portsall buoy seemed frightening close, blinking twice every nine seconds. A quickly plotted position showed that it was 1.3 miles away.

On a small vessel, anchoring is a positive, reliable manoeuvre. Provided enough chain is let out without actually fouling the anchor's flukes, it will usually dig in and bring a moving ship firmly to a halt. A coaster bringing up to a berth in a narrow river will often use her anchor to swing herself round in a single, well-controlled movement. But as the size of the ship gets bigger, so the relative size of her anchor gets smaller, until by the time one is

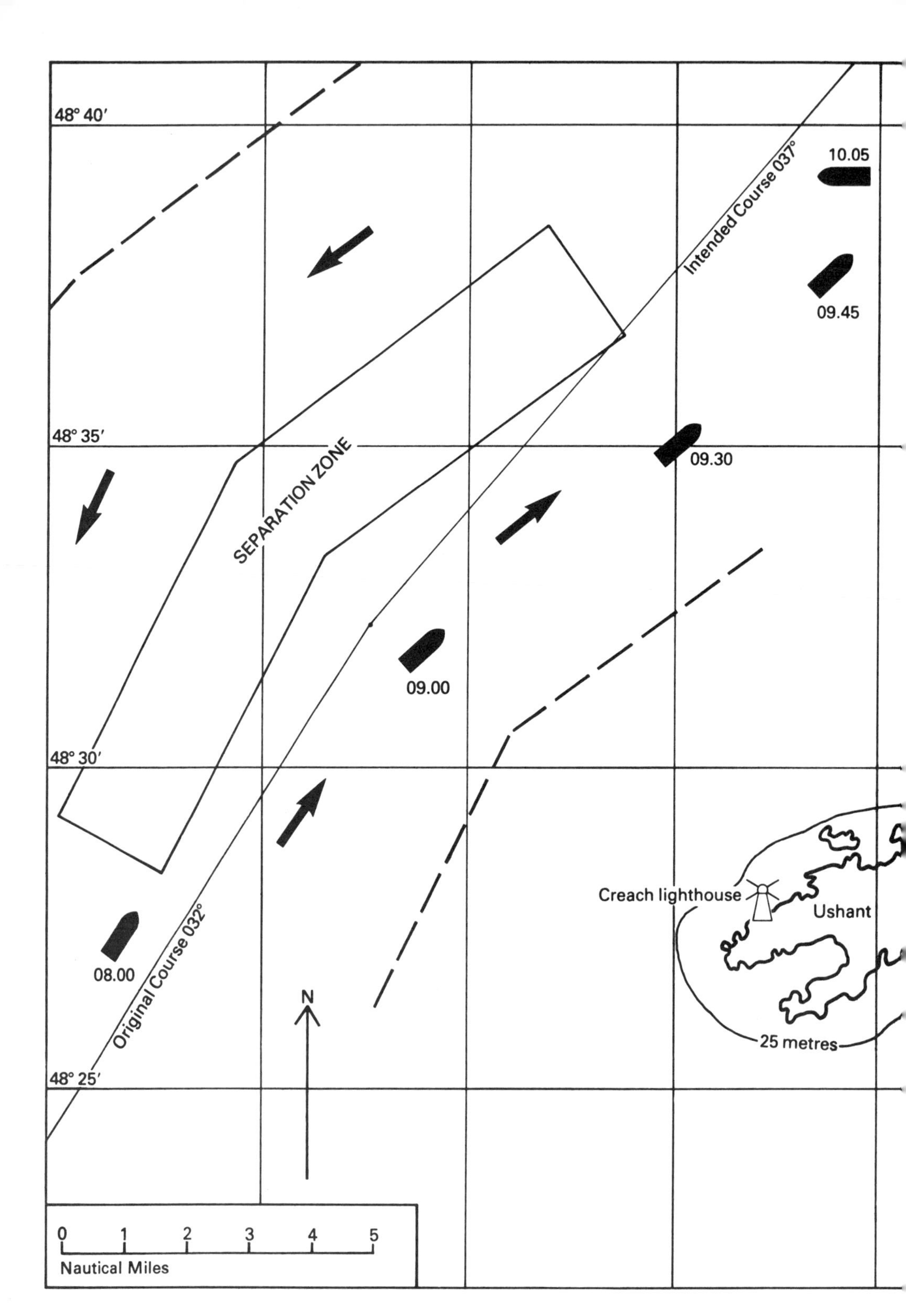

48° 40′
48° 35′
48° 30′
48° 25′
Intended Course 037°
10.05
09.45
09.30
09.00
08.00
SEPARATION ZONE
Original Course 032°
N
Creach lighthouse
Ushant
25 metres
0
1
2
3
4
5
Nautical Miles

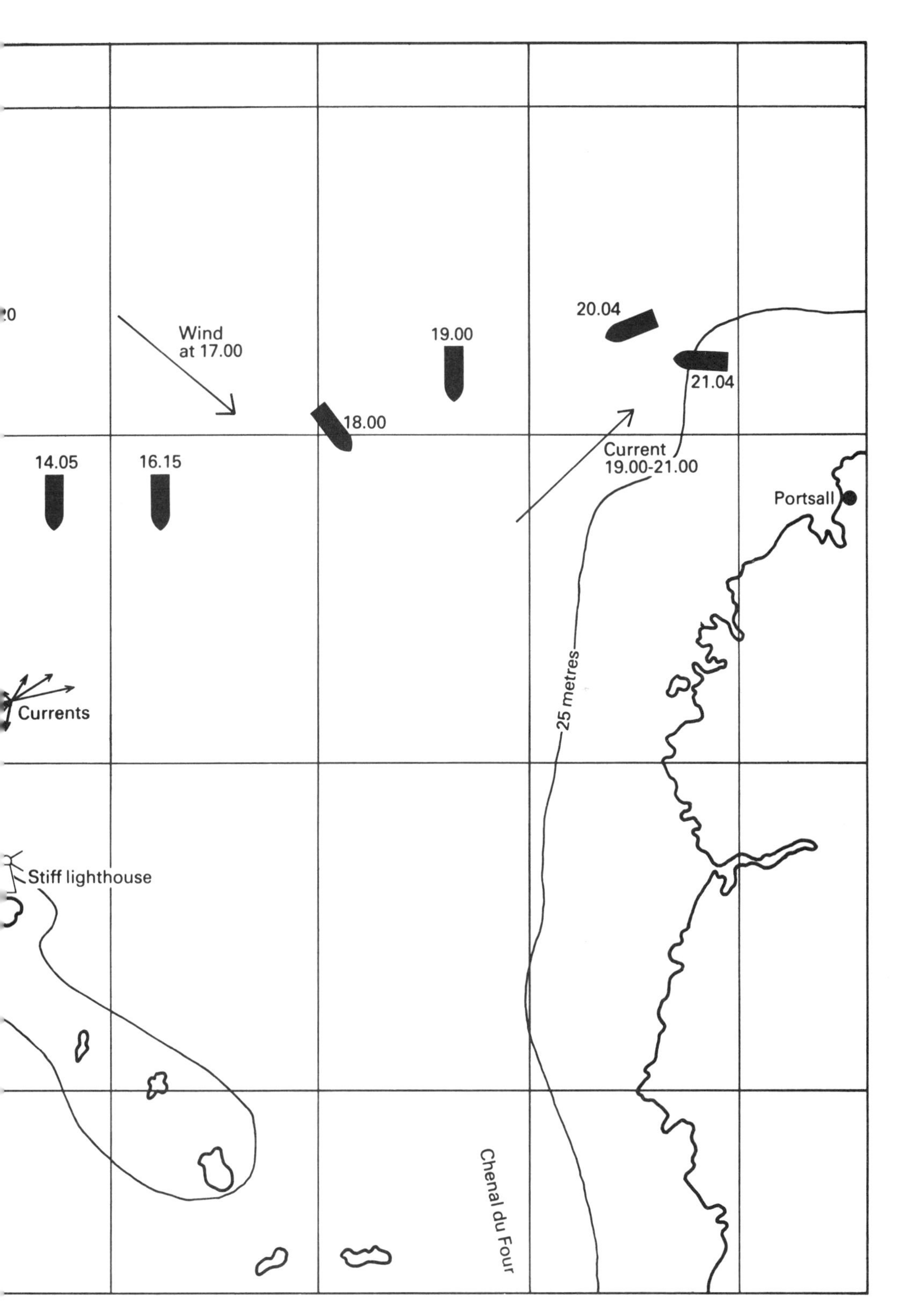

This diagram is based on the chart on which Captain Bardari plotted his ship's progress from the moment when her steering failed at 0945 on 16 March until 2104 when she grounded on the rocks off Portsall.

dealing with a supertanker anchoring is something to be contemplated only with great caution. The momentum of a 200,000 tonner is so enormous that to drop an anchor with more than one or two knots of speed on risks either breaking the anchor or plucking the winch clean out of the deck – either that or dragging the anchor uselessly through mud or sand.

So when Captain Bardari decided to let go the port anchor off Portsall, on to a rocky bottom with his ship driving before the gale at something like two knots in a vast and steepening swell, it was more in desperation than in hope. The heavy cable stretched taut, snatched, slackened and tautened again – presumably as the anchor snagged on a succession of sea-bed rocks. It was dragging, and when it was later salvaged the reason was clear enough. The flukes had been torn off. But even so it had some effect. It pulled the tanker's bow to starboard through another valuable ten degrees until she was heading almost due west.

It was the final irony of this long day that only now, with the rocks right under her lee and the tug committed to towing from the stern, the *Amoco Cadiz* turned her bows to seaward. If only she had turned a little earlier and the *Pacific* was there, ready to steer the bow; if only the rudder could after all have been fixed somewhere near amidships; the crippled tanker might have plunged off under her own power and gained a few vital miles of sea-room while she waited for the *Simson*, now just over forty miles away to the north and steaming towards her at fifteen knots. But two minutes after the anchor went down, it was too late to change the plan:

20.06 GUN LINE ACROSS

Both crews were committed.

Hampered by the wind, the darkness, and the unsynchronised rise and fall of the two ships, they hurried to complete the tedious process of passing the tow from one to the other – firing line, heaving line, rope messenger, wire messenger, and finally the heavy wire pennant and chain:

20.23 STARTING TO MAKE FAST WITH TOW AFT.

20.25 EIGHT SHACKLES PORT ANCHOR. GREAT STRAIN ON CABLE.

At this point, with the port anchor down and apparently slowing the ship's drift, it might have seemed sensible to let go the starboard one as well. Indeed there was some especially sharp cross-examination on this question at the London inquiry, including the suggestion that the starboard anchor winch may have been faulty. But Captain Bardari was adamant that with the ship lying as she was, it was too dangerous to work on that side of the forecastle, and in this he was supported by Mr Maynard:

'By this time the waves were coming in actually over the bow of the ship instead of over the side of the bow, actually over the front, and the water was

The wreck on 21 March (top) clearly showing how the ship's back was broken just forward of the bridge. As the *Amoco Cadiz* broke up her bows rose, pointing in towards Portsall two kilometres away and by 3 April (bottom) only the tip of the funnel and the point of the bow remained visible.

Top The heavily oiled beach at Roscoff; the streaks are caused by the oil running down the beach in the wake of the retreating tide. *Bottom* The worst affected area of saltmarsh was behind the Ile Grande.

pouring across the deck, she was rolling heavily, the water was swilling from side to side and I think we were lucky that we did not lose anyone as it was.'

Any debate there might have been was in any case cut short by yet another malevolently-timed stroke of misfortune:

20.28 BROKEN STEAM PIPE FORWARD. STEAM LEAK IN CAPSTAN STEAM DRUM.

The hiss of the escaping steam could be heard on the bridge. The Chief Engineer was sent forward to investigate but there was nothing to be done in those awful conditions. The steam cylinders had been blown clean off the port side of the winch. The pressure was gone, and now there was no power either to control an anchor cable as it paid out, with the help of the brake, or to haul it back in. The Italian captain ordered his men off the bow and turned his attention back to the stern, where another team were struggling to make fast the tow.

Chief Mate Strano and his men had managed to get the heavy wire aboard and take some turns on the bitts. It needed more turns to be really secure against slipping. Yet Bardari's voice could already be heard crackling over the walkie-talkie, urgently calling them in to shelter. Strano ran up to the bridge to ask him for another five minutes, but the captain would not listen. He knew from the chart, and the bright light of the buoy blinking alongside, that his ship might not have another five minutes. He ordered the mate to clear the stern immediately, and signalled the *Pacific*:

20.55 TUG INFORMED EIGHT TURNS ON BITTS, ANCHOR STILL DOWN. START PULLING EASY.

The German vessel acknowledged the message and responded with a request for the tanker's engines to be put 'Dead Slow Astern'. Third Mate Costagliola complied, but his hand had barely released the engine telegraph when Bardari – who had heard the radioed request from the adjacent chart room – dashed out and slammed the lever to Stop. He had no intention of starting his engines in any direction but ahead. As he saw it, the only help the tug could now give him was to pull his ship's stern further to port, so as to point her bows straight into the onshore gale:

21.00 TUG ADVISED PULL OUR STERN TO PORT. CAPTAIN TOLD THE TUG WE MUST GO AHEAD OR WE WILL GO AGROUND.

The Italian captain had never wavered in his belief that the tanker's best hope of escape lay in her own powerful engines, capable of driving her ahead at more than fifteen knots. But even now he hesitated. While the *Pacific* was still close under her stern there was a danger that the bigger vessel would heave the tug over as she moved off. Tugs had been sunk like that before. And in any case the *Pacific* seemed perversely intent on pulling to starboard, not to port as he wanted.

In the end, neither captain's judgement was put to the test. For the *Amoco Cadiz*, time had run out:

21.04 SHIP AGROUND AFT. ALL LIGHTS ORDERED OUT. PUMP ROOM FLOODED.

Even months afterwards, in the dry, formal atmosphere of the RINA's lecture hall, those who had been on the tanker's bridge were able to give a vivid account of that moment. With a lawyer's deliberate naivete, Dr Wiswall asked Mr Maynard how he knew the vessel had grounded:

> There was a great grinding crunch. There was a rush of air from the pump room fans and the pump room itself – the doors were blown open – and in about eight seconds the front of the bridge was covered in crude oil where the water and oil mixture coming up from the pump room flew straight up the front of the bridge. . . . I think the ship was ruptured by way of the pump room, that is the forward bulkhead of the pump room and the back end of Number 4 tank, and the pump room just simply flooded with oil and sea water, very very quickly . . . oil and water and air mixture came out of the fan head at the top of the pump room and blew straight up the front of the bridge.

The *Amoco Cadiz* had grounded almost exactly at high water – the worst possible moment from the point of view of later salvage because there would be no further rise of tide to lift her clear. But in fact it was an almost academic point in those conditions. She was not aground in fine weather on some estuarial sandbank, but grinding and twisting in an ambuscade of underwater rocks, at night, with the *Simson* still miles away to the north east.

Certainly Captain Bardari wasted no time on anything but the immediate problem of preventing his ship exploding in a great fireball as the mixture of hydrocarbon gas and air surged through engine room and accommodation. The crude oil she was carrying was of a relatively volatile type. Even before she struck he had telephoned round the various departments warning them to be ready to close down the machinery and switch off all the electricity on his command, for fear a spark would ignite an explosive gas mixture. Now he gave the order and within seconds the ship was dead.

Now there could no longer be any doubt about the need for an SOS call, the VHF radio was not available to send it. Instead, some of the crew sent out calls on portable radios, switching from channel to channel, while Maynard began searching for the least dangerous spot – where the smell of gas was least – from which to let off some distress flares. He found it right in the corner of the port bridge wing with the gale blasting clean across it, held his breath, and fired the red flares into the darkness.

21.23 ALL HANDS ACCOUNTED FOR.

21.30 SECOND GROUNDING AFT.

It was this second grounding which really finished the *Amoco Cadiz*, even though for days afterwards experts were talking optimistically about saving her or her cargo. It seems that when she first struck, she must have settled over an underwater pinnacle of rock, only to be lifted clear by the next big swell. Then she drifted on, dragging anchor and tug with her until she struck once more, this time in the engine room, right after. Again, Lesley Maynard's recollection was vivid:

> We knew the engine room was flooding because we could hear the air rushing out of the engine room fan heads. It was screaming in fact. The important bit that came up then – we were scared there was going to be a boiler explosion, because although the boilers had been turned off they were still hot, and in fact there was – there was a big bang and a cloud of flames came out of the top of the funnel, and to our amazement, nothing else. It just went quiet again.

21.45 PORT LIFEBOAT PREPARED – AND WASHED AWAY.

22.00 ENGINE ROOM FLOODED AND SHIP SETTLED AFT. POLLUTION OBSERVED. FLARES FIRED AT INTERVALS. ALL CREW ON BRIDGE.

In fact the phrase in Maynard's log about the lifeboat being washed away was written in later than the rest of the entry, after the tanker had begun to settle: 'A large wave came round the superstructure and filled the lifeboat up. It broke into three pieces. The centre section dropped into the sea and the two ends were left hanging on the port davit.'

All this time the *Pacific* had vainly been churning away at the stern, trying to pull the *Amoco Cadiz* off. But at 22.12 her own bridge log records the second tow parting. So now she surged clear and stood by, her searchlight trained on the darkened, helpless tanker while her crew recovered their damaged gear, radioed for shore assistance and wondered, no doubt, what would be left to salvage by morning. By then the *Simson* would be around to help, together with every other nearby salvage tug that got wind of the casualty.

At midnight, the first helicopter manoeuvred over the *Amoco Cadiz* to lift off the crew, starting with the injured man and the Chief Mate's wife.

01.45 ALL CREW LIFTED. CAPTAIN AND MAYNARD REMAIN.

A matter-of-fact record of a courageous decision. As Maynard recalled it: 'The Captain said to me, "When they go off, I am staying," and I said,

"Yes, OK I'll stay with you." The idea was that we were hoping to be able to give some assistance in the morning when daylight came. It was obvious that if we were still there someone would come and attempt to take the ship off – we thought.'

For the next two hours the two men sat in the wheelhouse gradually getting colder, watching and waiting. The only event Maynard could specifically recall later was how three pieces of paper, on which important notes had been made by the watchkeeping officers, came to be lost. It was a curious thing to remember in such detail, but it incidentally demonstrated for the London inquiry's benefit what enormous forces were at large as the giant tanker began to break herself over the rocks:

> They (the three pieces of paper) were given to me and because they were getting covered in crude oil with the other bit I had, I jammed them in the corner of the port bridge wing window, inside, and after we had been ashore for a while the lower doors of the accommodation got smashed open by the sea and the sea was coming up and down the inside of the ship at a tremendous rate, causing a violent wind to be sucked in and out of the bridge doors, and it blew the pieces of paper over the side.
>
> It also blew all the internal doors off their hinges because the captain and I went down and closed them, and jammed them closed, and they just ripped the frames out of the bulkheads.

By now both men were becoming badly gassed. They knew it, and tried to recover by taking spells in the relatively fresh air on the windward side of the bridge. But out there they had to contend with the cold wind and a drenching mixture of rain and oil, and they soon retreated to shelter again. This was the pattern until four o'clock that morning, when Maynard witnessed something that his fume-laden brain was at first unable to interpret:

> The captain was sitting inside the bridge and in fact I had been sitting inside with him. I went out on to the starboard bridge wing to try to get some fresh air, because I think we both realised at that stage we were very gassed, and I think we both thought that our chances of survival then were a bit slim, and it was while I was on the bridge wing I thought I saw some lightning flashes. My immediate reaction was that another squall was coming our way and that the weather was going to get even worse.
>
> 'I recall looking at the sky and thinking, "Where the hell is it?", because I could see stars all around and you cannot have stars all around and a squall. In fact, when I looked forward I could see what was causing it. In fact the ship was breaking in two, and as the metal was ripping it was flaring and flaming and

popping. So I called to the captain – I cannot remember the exact words – and said something like, "Here guv'nor, this thing's just fell in two." And he came out of the door very, very quickly with a lamp – a large, intrinsically safe damage control lamp – we shone it forward and we could see that the ship was broken, and breaking in fact right across.

'I think the captain said to me, "What are we going to do if this bit sinks?", and I planned to go off and try to get a life raft up on to the bridge, and when I looked round they had all gone.

'They had been washed away, as had the lifeboats. We had nothing left at all in fact, and so I fired our remaining three flares.'

The fact that the *Amoco Cadiz* was sinking must also have been obvious by now aboard the two Bugsier tugs, lying off with their big searchlights trained on her. The *Simson* had arrived at about 22.30. She sent a reassuring message, picked up on one of the walkie-talkie sets, promising to keep an eye on the Liberian vessel during the night, then close up in the morning to take soundings round her and endeavour to get a tow aboard. But the tanker was visibly settling by the stern.

An hour after firing the last of the flares, Maynard meticulously signed off his damp, oil-stained log and stuffed it into his pocket:

05.10 HELICOPTER RESCUES CAPTAIN AND MAYNARD.

From now on, what happened to their ship mattered less than the black tide oozing from her ruptured tanks.

7 / A Fatal Complacency

ASHORE, 16 MARCH had dawned cold and blustery, with gusts of wind and squalls of rain making the long, straight streets of Brest seem grey and dreary as people made their way to work past the still lit shop fronts, collars turned up and huddled into their coats for warmth.

Even before the office clerks and the shopgirls were up and about, the watch had been changed at the Navy operations room, known to everyone in the Brest naval base as COM, buried deep under the picturesque fort overlooking the harbour, which is naval headquarters.

The officers and men coming on watch had taken a last look at the restless sea and the heavy grey clouds scudding across the magnificent natural haven of the Rade of Brest, and, before entering the welcome warmth of the Marine Prefecture, had wondered when spring would arrive.

The *météo* reports were not good and the watch knew already that the local fishing fleets had decided not to venture out that day, and that the Navy patrol boat which made occasional tours of duty up and down the Channel navigation lanes had been kept in port. In fact no French warship was due to cross the area and they pitied the crew of any civilian ship that would find itself rounding Ushant in such heavy weather.

Out along the exposed Finisterre coast where the sea pounded heavily against the rocks, the observation posts which were the eyes and ears of the operations room were also setting up their day watch. The important Second Maritime Region, of which COM was the nerve centre, had twenty-three of them around its perimeter, making Brittany one of the best watched coastlines in France. Which was as it should have been since Brest was a naval arsenal and the home of the Ile Longue base for France's nuclear submarines.

Five of the posts watched the vital corner of Finisterre around Ushant and were to be involved in the day's events. One, a *vigie*, or look-out, kept a twenty-four hour visual, radar and radio watch from Pointe de St-Mathieu, a headland between the Rade de Brest and the string of small islands and reefs which straggle out to Ushant. And four *semaphores*, or signal posts, of which only two had radar, kept a daylight hours watch from points strung out along the coast further north. They were at Créac'h on the west of Ushant itself, and at Stiff on the east of the island (the two with radar); on the Ile de Molène between Ushant and the mainland; and at l'Aber Wrach, on the end of an inlet overlooking the vicious reefs off the fishing village of Portsall.

Other organisations involved in watching over those at sea were opening their own day. At Le Conquet, near to Pointe de St-Mathieu, the marine radio station Radio Conquet tested its powerful ship-to-shore transmitters and radio-telephone links after its night watch; especially those on which it would be maintaining a permanent listening watch for distress calls. And much further up the coast at Jobourg, near Cherbourg on the Cotentin Peninsula, the nearest of the French Merchant Marine's Regional Search and Rescue Operations Centres and the one responsible for the waters around Ushant, ran through its battery of radio, telex and telephone links.

France had four of these Search and Rescue Centres, permanently manned by Navy staff, either directly paid for on the military budget or credited to the Merchant Marine, to co-ordinate information in case of emergency at sea. The first, code-named CROSS from its French initials, was at Cap Gris Nez in the Dover Straits; the second, CROSSMA, at Jobourg; a third CROSSA was at Etel, near Lorient in southern Brittany to cover the Atlantic coast; and the fourth, CROSSME, at Toulon, for the Mediterranean. CROSS also included the sophisticated marine traffic navigational control centre by which the French operated their side of the Straits' obligatory shipping lanes; a similar control centre was in the process of being completed at Jobourg to cover the lanes off the treacherous Casquets (CROSSMA had just been equipped with a futuristic radar, already being used by CROSS, which enabled it to lock-on to individual vessels); and, disastrously late in view of what was about to unfold, funds had just been released for a third control for the Brittany area, to be built on Ushant.

If all these organisations had duties under French law making them responsible to one degree or another for policing shipping, protecting their coastline and saving human life, hard experience had also taught them to be realistic about their limitations in one of the busiest shipping areas of the world, covering several hundred square miles. Many of those on duty that morning had come to the same hopeful conclusion as they went about their tasks: with the weather the way it was, perhaps it would be a quiet day. Only those vessels which really had no choice would be at sea in such conditions. They tried not to concern themselves with the corollary of this, that the kind of wind that was now rocking the little lookout towers, never much less than Force 7 and gusting to Force 10, together with the huge rolling seas in turmoil on the offshore reefs, meant that any vessel which did get into trouble that day might expect nothing but the worst.

The *Amoco Cadiz* had arrived unannounced in their world, to the south west of Ushant, at 08.00 that morning. Captain Bardari saw no reason to make his presence known and certainly those on watch along the coast paid no particular attention to his approach. It was not until after the catastrophe that France demanded that such ships check in by radio as they entered French waters. If anyone saw her as she ploughed north, or plotted her route on a radar screen there was no official record of it. Nor would there be likely to be; for hundreds of ships enter and leave the English Channel every day and even in heavy weather the *Amoco Cadiz* might travel, unidentified, as just another part of the traffic.

And so it was the tug *Pacific* which first entered an official log on 16 March. At 08.30, putting out of Brest, she was spotted by the Brest naval lookout leaving the Rade and logged by COM. She turned north and used her powerful screws to head up through Chenal du Four between Ushant and the mainland.

But sometime after 09.00 came the first mysterious radio message hinting that some kind of drama was about to unfold at sea. The radio operator at the naval lookout at Stiff, on a routine listening watch, turned to VHF Channel 16 and suddenly caught a crackling and very faint message in English: 'Engine . . . keep clear, position 4836N 0503W.' The position corresponded to a point about eight nautical miles north of the observation post but masked by land. The operator did not hear the name of the ship making the broadcast, but in a busy shipping area such as that off Ushant, such calls, which are the sea-going equivalent of a roadside breakdown, could come practically daily. The ship either managed to effect its own repairs or went on to call for assistance. Hearing no more and despite the deteriorating weather, the observation post let the message go and did not inform COM.

The message must have come from the *Amoco Cadiz*. Though it was logged twenty minutes before 09.45, the time at which the French and Liberian inquiries were told helmsman Fede first discovered that the massive tanker would not answer to her wheel, the position given was about right and no other vessel in trouble emerged in those waters that day.

An hour later, an operator at Radio Conquet listening on 500khz had an almost similar experience. He heard the transmission of a TTT message (the first rung of a scale of messages of escalating urgency leading up to a full SOS) ending with the words '. . . keep wide berth, steering failure'. Again the name of the ship was not heard and there was no reference this time to engines. But it was almost certainly the first contact between the supertanker and those ashore describing her trouble. In a methodical manner, the radio operator pressed his transmission switch and began asking for a repeat of the message, broadcasting his request on all the major wavelengths used by commercial shipping. There was no reply. Not knowing who the ship was, or even where she was, the radio operator decided to take no action. The call was dutifully logged and forgotten.

It was almost immediately overtaken by one from the *Pacific*, checking in with Radio Conquet to make a telephone call to Brest, and announcing that the tug was moving up Channel. The station logged that call too in exactly the same way. But even while the tug and the radio station were speaking, the lookout post at l'Aber Wrach heard yet another unidentified broadcast on Channel 16. It was the first that particular post had heard that morning and it took slightly more action: it alerted all the other posts on that section of the coast and asked them to listen out.

But at 11.05 the *Amoco Cadiz* suddenly declared her presence in the area by calling Radio Conquet. Without any immediate reference to any earlier broadcasts she might have put out, her operator simply asked if he could make a long distance ship-to-shore telephone call on 2038 and 2726khz. The call, for Bardari, was intended to let his company headquarters in Chicago know about the battle which his engineers were even at that moment waging in the steering compartment below him. Miscalculating the time difference between Europe and America, he failed to raise Amoco and instead asked for calls to Genoa, in Italy, to try and reach two Italian captains he knew, Milanesio and Scarel, who had worked for Amoco. Though a captain is in command of his ship and must take responsibility for her, Bardari was reasonably young and was in a position in which even an experienced master might have wanted to take advice before making decisions. Again there was no reply and the tanker went off the air. Almost immediately though she was on again, and loud and clear on 500khz, she gave the first description of the drama to come: 'Need a tug assistance position ten miles north of Ushant steering gear failure.' And she added two ominous words: 'Full cargo.' While Radio Conquet was reacting to that, Bardari came on again and asked the station what tugs were in the area. The station told him the *Pacific*, of 10,000 hp, was near and as a service immediately re-broadcast the tanker's call for assistance to her. Whether Radio Conquet realised the full import of what it had just been told is not clear. But the station made a fateful decision: it logged the exchange of calls but kept them to itself. Neither COM in Brest nor CROSSMA at Jobourg, the naval and civilian operations centres, were alerted to the fact that one of the biggest supertankers in the world was broken down near to the coast, fully laden, in storm conditions, and that a single tug was her only immediate hope of aid.

Though the radio station could not see the *Amoco Cadiz* she was at that moment drifting slowly towards Ushant with her bows pointing south; and the *Pacific* was leaving the coast off Portsall Bay to the east, about fifteen miles away from her. At 11.28 the *Pacific* and the *Amoco Cadiz* spoke to each other directly for the first time. The first question from Captain Weinert on the tug was about the tanker's cargo: was she in ballast or fully laden. When she replied that she was carrying 223,000 tons of crude oil, Weinert instructed his operator to ask for Lloyds open form terms and swung his vessel around. The wind as the tug began to put back west was reaching sixty-seven knots.

The naval observation posts, with no knowledge of any of this, had stayed on radio alert, and that at Créac'h suddenly heard the *Pacific* on Channel 16, speaking in English. Still none of the stations had seen or identified the ship whose messages they had earlier intercepted, and the lookout casually kept his radio open to see if he could hear a position from the exchange with the *Pacific*. It was clear that the tug, because she was speaking in English, might have raised the mystery vessel. He understood only the words 'north Ushant' and a visual check out to sea showed nothing in his sector. Out of curiosity the lookout called Pointe de St-Mathieu, which kept a radio watch on other frequencies, but there was no information forthcoming and the lookout went back to other duties.

Meanwhile, Bardari, unhappy with Weinert's insistence on open form salvage rather than a towage contract for his ship, re-opened his contact with Radio Conquet and asked again for his call to Chicago. He apparently passed no further information on the plight of his ship to the radio station, despite the fact that he had now been told that his steering was damaged irreparably, nor was he asked for any. This time the call went through and he explained to Chicago in detail. The conversation, which, by law, Radio Conquet was forbidden from monitoring, lasted for fourteen minutes. While the Italian captain was speaking, Weinert put a call through to Hamburg and briefed the Bugsier company in a terse three minutes. Bugsier made the decision to send him help, in the shape of the *Simson*.

For half an hour radio silence fell so far as the *Amoco Cadiz* and anyone listening out for her on shore were concerned. And then at 12.18 a telephone call for the tanker came through the Radio Conquet, routed via Brest. In between its other radio work for shipping, the station tried to contact the *Amoco Cadiz* on all wavelengths for at least fifty minutes, but she failed to reply. This, the governmental inquiries were told later, was the crucial period in which the two captains first argued about the terms of their contract, ending with the realistic decision to get a line aboard and argue the legalities later.

For at 13.16, the signal post at Stiff, having finally raised the *Pacific* on Channel 16 and asked her what was going on, was told shortly by the tug that she was towing the Liberian tanker *Amoco Cadiz*. Stiff, who seemed unaware of the tanker's size but realised that, though no one could see her, she must be very close, called COM in Brest. It was the first call that day alerting the Navy to the fact that something unusual was happening at sea.

Almost immediately, the tanker suddenly appeared through the murk from behind Stiff lighthouse, which had been masking her from the observation post. Even six miles away, wallowing in the heavy, wind-whipped seas, the ship should have looked enormous. The lookout took a bearing and transmitted it to COM via Pointe de St-Mathieu.

This message was one of the most important made that day. For it set COM's immediate attitude to what was happening. What, the operations room might have asked, was the tanker doing drifting out of the navigation lanes and in trouble only a few miles from a lee shore. But the message from Stiff read: 'Azimuth 012 from Stiff, distance six nautical miles, route SW, 6.5 knots.' Though the distance may have been accurate, the heading conflicts with the evidence of the tanker's crew and the speed, even allowing for the wind and tide, was clearly impossible. What it suggested, in a dangerously reassuring manner, was that the tanker was somehow under way at a respectable speed, clear of the coastal reefs and with a line aboard her from a tug, and that she was a hazard neither to herself or to the coastline. The stand-in duty officer at COM read the flimsy, and in the words of the French Senate investigation later 'did not judge the information alarming and did not react'. No one, it seems, checked on the sheer logistics of what was happening and asked themselves if it could be done.

Nevertheless the Stiff lookout decided to keep a watch on the route of the two ships and to note down their positions every half hour. As Stiff began its task, Radio Conquet alerted the *Pacific* that the second Bugsier tug, the *Simson*, was heading towards the scene as fast as weather would allow and was just passing Cherbourg: her estimated time of arrival was 23.00 hours.

Though the *Pacific* acknowledged the call, Weinert's mind was on other things, for unknown to those watching out for the ships from shore, the argument over salvage terms had broken out all over again. And Weinert knew that his tug was in no position to tow the *Amoco Cadiz* anywhere. In his mind he already knew that the battle now was going to be simply to keep her off the rocks. He was going to need the *Simson*.

At 15.00 the lookout at Molène caught sight of the *Amoco Cadiz* for the first time and contacted the Créac'h post on Ushant to ask for information. Créac'h must have been talking to Stiff, for even though it was no longer in a position to see what was happening, it replied reassuringly, 'Tanker under tow. No need to worry yourself.' None of the posts seemed at all worried by the tanker's route and even Stiff, which had been tracking it did not seem concerned enough to contact COM again.

The *Pacific* called Hamburg again via Radio Conquet and for ten minutes explained Bardari's position over the contract. Then she spoke directly again to the tanker at 15.15 to plead with the captain to change his mind over the terms. Bardari reluctantly agreed, obtained approval at 15.45 when he eventually spoke to Chicago, and then informed Weinert ten minutes later. That, at least, was the evidence of the two ships' captains.

But the log of Radio Conquet, submitted to the French inquiry, shows a slightly different version of events. At 15.36 Weinert dictated to Radio Conquet for re-transmission to the *Amoco Cadiz* his formal telegram proposing open form conditions to Bardari. No 15.45 call to or from Chicago was logged. But according to the French at 16.00 hours the tanker radioed the tug accepting the proposal.

The weather was worsening again and the wind, blowing now more from the west was gusting up to sixty knots. Then at 16.15 came one of the dramatic turning points of the day as the tow line between the tug and the tanker parted. The inquest aboard both vessels as to what had gone wrong must have still been in progress when at 16.28 the radio operator on the *Amoco Cadiz* contacted Radio Conquet with the tanker's telegram acceptance of Lloyd's open form to be re-transmitted to the *Pacific*. The tanker churned its engines astern to try to keep clear of the reefs now just over six miles away. And then, as the crew of the *Pacific* began to make ready another line, the *Amoco Cadiz* started to drift away first east and then north east, parallel with the coast. Weinert spent his time, as his crew worked, in a flurry of telephone calls to Hamburg. Bardari, waiting, spoke just once, briefly, to Chicago. The two ships were now in some of the worst weather conditions of the day for the wind, still very strong, had turned cross current and the huge waves measured twenty-five feet from crest to trough.

Then at 18.30 an incredible thing happened in view of what was to come. With the tanker and tug still battling to reconnect the broken tow in the falling dusk, the naval lookouts at Stiff, Créac'h, l'Aber Wrach and Molène closed down their watch. They had reached official sunset.

Now the only official station on duty watch was the radar-equipped *vigie* at Pointe de St-Mathieu. But as the two ships moved further to the east they were in any case moving gradually out of even radar range of this station, positioned as it was close to the tip of the Rade de Brest. Thus, even if COM had decided it had wanted accurate reports on what was happening to the stricken tanker, no one was any longer really near enough to give the information. But COM had not asked for such reports. No attempt was made to persuade the watch to stay at their posts and keep an eye on the two ships until it was too dark for them to see any more. Despite what was happening at sea an unaware French Navy had so far treated 16 March as if it were a normal day.

It was almost an hour and a half later, at 19.56 that Jean Gouzien's call came indicating that disaster was imminent. But it was not made to the Navy. The deputy skipper of the Portsall lifeboat had walked down through the little port towards the lifeboat house on a spit of raised ground from which, with the wind full in his face, he could look out to sea across the ragged reef of the Roches de Portsall, two miles out, to the Corn Carhai light beacon. Gouzien saw the beacon, but he also saw something else. There, seemingly very close to the seaward side of the Men Goulven rocks, was a set of lights. In the weather conditions, Gouzien could not make out clearly what they were, but they appeared to him to be the lights of a ship, perhaps two.

He hurried to the telephone and called Radio Conquet, presumably on the basis that any ship that close to the rocks and in trouble would have sent out some kind of distress message. He reported 'a light about two miles off the rocks at Portsall'. The radio station received the news calmly enough, and though full details of the conversation are not known, it appears they did not automatically connect the sighting with the *Amoco Cadiz*. Though they had been handling calls to and from the tanker and the *Pacific* all day, they did not know exactly where the two ships were now. Also the seas around Ushant were a normally busy shipping area and even on this bad day there might have been other vessels in the area of which they would not have been aware unless they had had radio contact or had been overheard broadcasting. None of the messages or requests for telephone connections they had received from either the tanker or the tug had suggested any immediate danger. And so those in charge decided to act strictly according to the book: they called the CROSSMA regional search and rescue centre at Jobourg.

It was the first indication that CROSSMA had had that day that any kind of drama might have been taking place at sea. Though the *Amoco Cadiz* had lost its steering at 09.45 that morning, this first call to CROSSMA was logged at 20.00 hours, in the middle of the evening. Full details of the call are not known, but it appears that the names of the two ships were not mentioned.

In answer to the Radio Conquet call, CROSSMA replied: 'OK. Receiving well. But these lights, are they stationary or are they distress flares?' Radio Conquet replied: 'No, not distress flares but the two ships are very near the rocks.' CROSSMA said, 'OK, they are navigation lights.' And the radio station then gave CROSSMA Gouzien's telephone number so that they could check directly with him.

But in the meantime Gouzien had been doing some checking of his own. With no knowledge of the day's events and not wanting to be alarmist, he had set off along the coast to see if he could get a better vantage point from which to see the mysterious lights. From the top of the headland he got a view which apparently satisfied him, for at 20.10 he called CROSSMA and told them he had sounded a false alarm. It was, he reported, 'a tanker towed by the *Pacific*'. The striking silhouette of the huge tug would have been known to Gouzien, even if he had not been alerted to her presence, for she had been based at Brest for some time. CROSSMA called back Radio Conquet and to put the operator's mind at rest said simply that Gouzien had reported 'after a patrol of the coast, that it was one boat pulled by another'. The radio operator thanked CROSSMA and went back to his duties without taking any further action.

But Gouzien was not the only person to be worried by the sight of lights so close to the shore. At 20.34 an anonymous caller telephoned the Pointe de St-Mathieu station, which was keeping the night watch, and reported that a convoy made up of a tanker and a tug was immobile to the west of Corn Carhai. Pointe de St-Mathieu immediately passed the information to COM. And a minute later another sighting was made. Joseph Callac, a lookout at the Molène signal post had not gone straight off duty when his watch ended at 18.30. Now, two hours later he had just clambered down the steps from the lookout room and was standing at the foot of the signal tower, looking out to sea. Incredulously, he suddenly spotted navigation lights off to the north east, between himself and the coastal reefs. He raced back up to the lookout room and took a bearing. The lights appeared to be almost on top of the rocks. Not knowing what the vessel was, but with an awful sinking feeling that it could well have been the *Amoco Cadiz*, he too called COM.

At sea the *Amoco Cadiz* was in the eleventh hour of its crisis in both senses of the phrase; the second line from the *Pacific* was now aboard her, but the tug could not stop her drifting. She had her port anchor down in a desperate effort to hold her position until the *Simson* might arrive. If the *Simson* and the *Pacific* could turn her head straight out to sea, she might still use her powerful engines to drive herself clear of the coast.

This time the naval operations room in Brest began to move. At 20.40 it put the Stiff and l'Aber Wrach lookout posts back on watch and began to call in its own off-duty staff who were going to be needed for any emergency. A check showed that the only naval vessel which might be of any use, the 4,600 hp tug *Malabar,* was on a survey trip to the Gulf of Gascony. COM radioed the vessel and discovered through a poor and fading contact that she was in the Groix Straits, near Lorient, or about six hours away. COM asked the duty officer at the naval base at Lorient to order her back north with all speed. And then the duty officer at COM called his own deputy operations chief. This was going to be a bad day for a catastrophe. The *Préfet Maritime* and Atlantic Zone Commander, Vice-Admiral Jacques Coulondres, responsible for control of French waters from Cherbourg to the Spanish border, had gone to Paris to take part in an official ceremony.

Urgently, the coastal observation stations tried to get in touch with either the *Pacific* or the *Amoco Cadiz* to find out what was going on. Stiff raised the small French tanker *Port Briac* which reported it had sailed past the two ships at 20.00 when they seemed just off Corn Carhai. Though no one ashore knew it then, disaster was now only minutes away. COM contacted Radio Conquet to see what the radio station knew. The station explained that it knew the *Pacific* was trying to tow the *Amoco Cadiz,* but that all calls had been normal and neither vessel had sent a message to say the tanker was in difficulty. Indeed, as this conversation was taking place, the *Pacific* came on the air again asking for a normal telephone connection to its Hamburg office. Radio Conquet complied apparently without comment.

At 21.00 the *Malabar* radioed to say that she was heading north as fast as she could go but that the weather was making it heavy going. COM thanked her and the duty officer stared gloomily at his dispositions list which showed that the only other two ships he badly needed, the *Malabar*'s two sisters, were unavailable. One, the *Centaure,* was in dock for an engine overhaul, and the other, the *Tenace,* on a mission to Newfoundland. No one at COM knew exactly how desperate the situation was, but as the hunt for the position of the supertanker went on, l'Aber Wrach lookout contacted COM to say it thought it could see her and reported bearings which put her very close to the rocks. Only a minute later the station called again with the devastating message that both tug and tanker had crossed a line putting them only a mile from the Men Goulven reef.

COM immediately called the Navy patrol ship *Chevreuil* and ordered her to put to sea. But it was already too late. At 21.04 the *Amoco Cadiz* had run aground for the first time. Twenty-five minutes later l'Aber Wrach, watching in the darkness, reported that the tanker might have broken her tow and made a half turn, but it was simply the ship's death throes.

Unbelievably now, there was still no general distress call from the supertanker. Even as l'Aber Wrach first spotted a red flare fired from the *Amoco Cadiz* at 21.42, Weinert was in the middle of a telephone call to Hamburg, telling Bugsier the news first. It was not until 21.50, nearly an hour after the first grounding that the *Pacific* sent a message to Radio Conquet. Still it was not a formal SOS, it read: 'We need helicopter for *Amoco Cadiz*, he shoots redlights on ground position 48.36.5N, 04.46W. *Amoco Cadiz* is aground Basse de Portsall.' But it made its point.

COM, told of the flare by l'Aber Wrach, then got itself involved in a communication mix up. Instead of calling CROSSMA at Jobourg, the search and rescue centre covering the area, it alerted CROSSA at Étel. CROSSA, realising what had happened, passed on a message to CROSSMA, which was the first full explanation CROSSMA had had of what might now be taking place at sea. But even then it was not entirely complete. It said that 'a tanker of 230,000 tons' was on the point of grounding in the region of l'Aber Wrach. CROSSMA called COM in alarm and demanded confirmation. COM told it that the 'tanker *Amoco Cadiz* of 230,000 tons under tow had broken its tow' and added as a guesstimate that the only help the navy could give, the *Malabar*, was about four hours away. COM also sent over to CROSSMA the message the *Pacific* had sent out and alerted the naval helicopter base at Lanvéoc – Poulmic.

Still trying to find out exactly what was happening aboard the tanker, l'Aber Wrach asked Radio Conquet to try to get through to her, but the radio station replied that it had tried without success. But almost immediately, at 22.00, the *Pacific* came on the air asking Radio Conquet for a call to Chicago which it was going to relay to the *Amoco Cadiz* on VHF Channel 6. The official French Senate report guessed it was done this way because of poor radio communications but though there had been trouble during the day, the fact was that by this time the tanker's main radio had already been shut down because of the fear of an explosion aboard, and Channel 6 on a lifeboat radio was the only communication it had left. The call went through with Chicago promising to call back in thirty minutes. Again the call seems to have been unmonitored, not was there any evidence that Radio Conquet used the communications opportunity to interrogate either ship.

But *Amoco Cadiz* then belatedly broadcast her own version of what was going on aboard. Even now it was not an official SOS. It reported to Radio Conquet simply that she was in 'Position 48.36N 04.43W loosing crude oil *Simson* is also near in distress position'. The garbled form of the message was taken to be owing to the tanker crew's bad English and miscomprehension by the French operator. But it was quite clear now that there was going to be a disaster at sea, with a rescue operation to be carried out in high winds and heavy running swell, and that the Brittany coastline was once more going to be covered with oil. Just how badly no one that night fully realised, but from the sheer size of the tanker's load alone things were obviously going to be grim.

When COM heard the repeat of the tanker's message it immediately gave instructions to put anti-pollution gear aboard the *Chevreuil*, which had been ordered to stand by in harbour, and ordered the l'Aber Wrach lifeboat to sea to back up the Lanvéoc helicopter crews who were making ready their machines. Radio Conquet reacted too and sent a telex to CROSSMA: 'The tanker has touched the rocks at Portsall, position 48.36N 04.46W. Request helicopters. Fear of pollution.' CROSSMA, after checking with COM, replied, 'OK. COM has been alerted and will rush planes and helicopters.'

Now that pollution was an added factor in the drama, the official network of those who would be needed to be involved began to spread. COM had to alert the central Navy operations room in Paris. The *gendarmerie* in Ploudalmézeau, a small market town just inland from Portsall, was given the job of finding the chief administrator of Brest at home and telling him what had happened. The man reacted immediately: asking whether CROSSMA had been informed, he got into his car and drove to the Marine Prefecture in Brest to begin what was going to be a long night of preparations for the fight against the oil which would begin at dawn. At the Brest naval headquarters itself, COM contacted Vice-Admiral Robin, who was standing in for the absent Coulondres, and told him what was happening. The local civil defence chiefs were alerted and contacted CROSSMA for information. And they were followed by officials from the office of François Bourgin, the Prefect of Finisterre, in Quimper.

Amazingly, at 22.45 the French subsidiary of Shell, the company which owned the oil aboard the *Amoco Cadiz*, which must have been told of what was happening by Amoco in Chicago, also contacted CROSSMA, told them of their involvement and that the cargo was Middle East crude, and offered any help the company could give.

All this time, more red flares had been seen from the *Amoco Cadiz*, the l'Aber Wrach lifeboat was getting near and a lone fishing boat, which had been in the area on its way home to Portsall was now standing by the tanker in case it could help. But there still was apparently no SOS call from either the tanker or the *Pacific*.

At 22.30 CROSSMA made check calls passing on everything it knew to

COM at Brest, to the Marine Affairs Department at Nantes and to its local Brest office, to the prefecture at Quimper, and to the leaders of the local anti-pollution squads. And even as it was doing so, Shell came back on again, suggesting it supply CROSSMA with a list of smaller tankers which might be used to lighten the load of the *Amoco Cadiz*, either to get her off the rocks or to empty her.

It was not until 23.18, nearly two and a quarter hours after she first ran aground, that an SOS from the *Amoco Cadiz* was heard, and then it came in a garbled form. Broadcast on 500khz it read: 'Tanker aground need immediate assistance — AS — Please AS — SOS QTH 4836N 0459W — tanker aground need immediate assistance if possible helicopter we have only Channel 6 VHF.' Radio Conquet re-transmitted the message to CROSSMA and asked if helicopters had been sent. Assured they had, the station then broadcast the tanker's message to all shipping on 500khz.

CROSSMA and COM held a hurried consultation. Though the naval Super Frelon helicopters were even at that moment getting ready to take off from Lanvéoc, the two operations rooms were worried about whether they would be able to do the job in the driving rain and gusting winds. The last weather report had shown a wind west south west at twenty-two knots gusting up to twenty-eight knots, far from ideal for a winch line operation. As the first Super Frelon went off at 23.30 they ordered the Ushant and Molène lifeboats to sea.

As they waited, the ever efficient Shell France came on again offering its list of tankers which would be available and a check list of available stocks of chemical dispersant which might have to be used on the oil. The call was a timely one, for the next picked up by CROSSMA was to say that the civil authorities along the coast had decided to launch their emergency anti-pollution plan, POLMAR. And Vice-Admiral Robin was to make up his mind within the next few minutes to give the go-ahead for the naval version of the same operation, code-named POLMAR-Mer. Pointedly, these messages were followed by another distress message from the *Amoco Cadiz*. 'Grounded full loading with pollution,' it warned ominously, 'require immediate assistance and crew rescued by helicopters.'

The helicopters were on the way. The first Super Frelon of flight 32F arrived out of the darkness over the wreck at 23.49 to a scene like a devil's cauldron. The tanker, in darkness itself, was being lit eerily by the pitching searchlight beams of the *Pacific*, and of the *Simson* which had now arrived, tragically too late to be of help. The ship, surrounded by an ocean already dark and menacing was now adding the contents of its own ruptured tanks to the water heaving in great seas around the bridge island and crew quarters astern, the oil splashing up on to the white paintwork and showering across the helicopter as pilot Lieutenant Jean Martin lowered his machine over the open deck. Six minutes later the first of the forty-two crew and the wife of Rosario Strano were being raised as Martin battled to keep the Super Frelon in position and prevent the cable from swinging wildly. In the darkness and the atrocious weather, the pilots could hardly even see what was below them, much less manoeuvre, and the official Government reports on the affair later devoted whole sections to congratulating the 'remarkable performance and courage' of Martin, his fellow pilots and crewmen. Indeed it was a heroic effort: Martin, with 5,300 hours flying time, 400 of them at night, was only one of five pilots in his squadron qualified for night flying, and even he only found the crew, huddled on the port wing of the bridge by flying down the searchlight beams from the tugs. Normal procedures were for the helicopter to stay at ninety feet while the crewman went down on the line bringing up a man every two to three minutes: but Martin took his machine down to forty-five feet and in forty-three minutes took on board Mrs Strano and twenty-seven members of the crew. Holding a large helicopter stationary for that length of time and in those weather conditions is almost unheard of.

At the stroke of midnight, the *Amoco Cadiz* acknowledged the work being done by the helicopters: 'Now helicopter is here to rescue thanks helicopter is here to effecting crew rescue here closing and going up on bridge standing by to abandoning the vessel.' The message was apparently not sent by Bardari or the English safety officer, Maynard, who had both made a decision to stay aboard, but it was the last radio message received ashore from the tanker.

The hazardous conditions meant that the helicopters following Martin had to work with precision and so it was an hour and a quarter later, at 01.15 that the *Pacific* radioed a progress report which also confirmed the French authorities' worst fears: 'About *Amoco Cadiz* now fifteen or twenty persons are still on board. She makes loosing oil pollution.' But by 01.47 of the new day all forty-two crew had been lifted off with the exception of Bardari and Maynard alone on the bridge. COM passed the helicopters' 'mission completed' message to CROSSMA, which then recalled the Ushant and Molène lifeboats, leaving that from l'Aber Wrach standing by to save the lives of the two men still aboard if that should prove necessary.

But the sea was not going to let the supertanker rest. The strength of the angry Atlantic, blackened by the oil bubbling from her tanks, began to shift her as she lay squatting helplessly on the granite peaks of the reef: pulling, twisting, ripping and buckling the sheet steel and tearing the *Amoco Cadiz* in two. Maynard described dramatically at the Liberian Government inquiry in London the tremendous showers of sparks and lightning-like flashes the tortured metal produced as the sea went about its work; and the sight must have been every bit as awesome from the small flotilla of vessels standing off.

The *Pacific* and the *Simson* remained, highlighting the scene with their searchlights, as did the lifeboat and the local fishing boat, and now a group of little anti-pollution vessels had arrived from Brest to begin tackling the spilling oil at first light.

Maynard went out on to the wing of the bridge and fired off the red flare that said he and Bardari at last wanted to be taken off. The lifeboat saw it stream into the darkness and radioed to shore. At 04.36 a Super Frelon which had been kept on stand-by took off to answer the call. And it arrived not a moment too soon: even as it approached the wreck, Maynard was firing off his last two flares, for the sea had reached the bridge. At 05.03 the helicopter began lowering its line and a few minutes later one of the world's biggest tankers was finally left alone with the ocean.

Around her the flotilla of ships and boats which would have work to do in the coming days was still growing. The l'Aber Wrach lifeboat had been recalled but replaced with the Navy patrol ship *Chevreuil* and her anti-pollution gear, together with the supply barge *La Libbelule* and the gunboat *Jean Moulin*. And almost at the last, at 07.30, as daylight spread over the sky, came the *Malabar*, the only naval vessel whose presence might have had the slightest effect on the events of 16 March.

Asleep in his home near Le Conquet and over eighteen miles from the grounded tanker, a government marine biologist, Dr Lucien Laubier, was wakened by his wife. 'Lucien,' she said, 'you will have to get up and do something about the central heating. There's such a terrible smell of oil all round the house.' Laubier got up and disappeared from his bedroom. In a few minutes he was back. 'It's not the heating,' he said, 'It's coming from outside. . . .'

8/Plan POLMAR

On the morning of Friday 17 March, the *Amoco Cadiz* lay abandoned in the centre of a ragged lake of oil four miles across, her back broken just forward of her bridge. The oil, heaving on turbulent waters being moved by a Force 6 wind from the north north west, reached from the reef on which she had impaled herself on to the rocks and coves of the shoreline and deep into the harbour of the little fishing village of Portsall. And more was cascading from her ruptured hull at every moment. By evening it would be ten miles south along the coast towards the small port of Le Conquet, and six miles east, licking at the entrances to the rich oyster growing inlets of Aber Benoit and l'Aber Wrach.

Even while it was still dark on that first morning the French had begun talking about what to do about her. They had had accidents with tankers before, too many, but none like this. Already they knew from the rescued crew that the spill was going to be an enormous one: the tanks where she had broken might release somewhere between 58,000 and 80,000 tons of oil. But the crucial question was what would happen next. The thought of all 223,000 tons ending up in the sea was almost too appalling to contemplate. And yet the ocean along the Atlantic coast of Brittany was an almighty force, particularly in the storm conditions now whirling around the *Amoco Cadiz*, and nothing could be taken for granted.

The weapon the French were going to have to use against the oil was a national oil pollution contingency plan they called POLMAR. First drawn up in December 1970 in the light of the *Torrey Canyon* disaster it was now deployed with a great deal of hesitancy. To begin with it had never been envisaged as having to tackle a spill of more than 30,000 tonnes, equivalent to the oil slick that had been washed ashore from the *Torrey Canyon*, or one tank of a modern supertanker like the *Amoco Cadiz*. The target had been set in 1972 and never updated and it was not entirely certain that the means were sufficient even for this. It had never been tried out on a major disaster and never been realistically tested either on land or sea. Even more disturbing, a new version of the plan had been written within the past three months which those who would have to work it had hardly assimilated, much less had a chance to put into practice. But it was all they had, and there was a strong feeling even in the first hours that no country had ever experienced anything like the battle they were about to take on with the *Amoco Cadiz*.

In the original plan POLMAR, general strategy, the recommendation of products and equipment and a programme for supplying them was in the hands of a body called CICOPH (*Commission interministerielle de lutte contre les pollutions par hydrocarbures*), chaired by the head of civil defence. CICOPH, made up of the operations chiefs of all interested ministries, could also call on expert advice from advisory bodies such as the French Institute of Petroleum (IFP), the National Centre for the Exploitation of the Oceans (CNEXO), and the Scientific and Technical Institute for Marine Fisheries (ISTPM).

Civil defence, through its permanent operations room, CODiSC, at Levallois-Perret, Paris, would act as a national liaison headquarters; providing meeting facilities for CICOPH, collecting and distributing information; keeping a check on the spread of any pollution being tackled, and the means used against it.

Regionally the means would be co-ordinated by a top civil servant in charge of one of four defence zones. And the fight on the ground would be co-ordinated between the *Préfet Maritime*, or commander, of the Naval region, where the oil was at sea; and the *Préfet*, or governor, of the affected administrative department when the oil came ashore.

The prefects were supposed to operate on the basis of a pre-established local plan drawn up by a committee chaired by the head of the defence zone and including the *Préfet Maritime*; the General commanding the local military region, and representatives of the various government authorities. The prefect of the department could be included if the defence zone chief felt it necessary. The decision on when to launch POLMAR was left in the hands of the Prime Minister.

But no one in any of these organisations had been nominated overall commander of the POLMAR operation; each organisation, and there were potentially up to eighteen of them, kept to its own self-imposed or allotted tasks and command structure, with orders merely to liaise and co-ordinate with the next at appropriate levels. Four ministries seemed central to the plan at a logistical level, but a close look at their duties revealed inherent duplications and contradictions. Defence would play a major role. At sea, through the Navy, it would be responsible for action over the polluting ship itself; it would send in its own ships to fight the oil and plan the requisitioning of any extra ships and boats needed; it would centralise stocks of materials for the fight at sea; it would collect and distribute information on the spread of the slick. On land, through the Army, it would be a major source of men and transport; it would operate special squads to help other ministries with specific tasks; and as the pollution spread it would help maintain co-ordination between task forces.

The Transport Ministry, which controlled the Merchant Marine and the Department of Marine Affairs, would have important duties in choosing the products to fight the oil at sea; requisitioning ships and boats; helping lay barrages on the coast; centralising the buying of products and equipment; keeping track of products as they were used; and, of crucial importance, organising the disposal and destruction of oil and debris recovered from sea and coast. The Ministry of Equipment (public works) would be in charge of reinforcing other departments with men and materials; aiding Transport with the stockpiling; removing and destroying oil and debris; and advising the Interior Ministry on pollution fighting techniques. Interior would have the job of assembling information on the disaster and its consequences; getting technical aid from experts; taking police measures; and helping local authorities with personnel and materials.

The end result of all this, concluded the National Assembly commission of inquiry which looked at the affair later, was an impressive document backed by confusion. Several ministries seemed to be in charge of the same tasks, and those given specific tasks often did not have the powers or means to carry them out. Affected areas were arbitrarily divided, for instance, into land, water near the coast, territorial waters and high seas, and responsibility for each given to a different organisation.

A perfect example was the provision of floating barrages. There were to be three applications: those for the open sea, those to protect bays and estuaries, and those for ports and shellfish beds. The Navy was in charge of the first, but it had to help Marine Affairs and the Ministry of Equipment with the second. The third was in the charge of Marine Affairs but in collaboration with Equipment, and in certain circumstances the Lighthouse Service, part of the Directorate of Ports and Navigable Waterways, might have to help. Beyond that it was up to Marine Affairs to buy all the barrages and Equipment to provide the necessary accessories to hold them in place.

Even the funding of POLMAR was complicated. For the Navy the bills were to be sent to the Ministry of Defence. But for operations ashore they passed through the prefect of the department to the appropriate ministry. After checking, each ministry would send its accounts to the Interior Ministry, as controller of Civil Defence, where they would be checked again and passed to the Ministry of Economics and Finance for payment. But all this would happen at the end of the affair and there would be a heavy administrative burden for departments which were supposed in any case to put any spending over £12,000 out to tender.

In 1977 an inter-ministerial group with special responsibilities for the sea called GICAMA (*Groupe interministeriel de co-ordination de l'action en mer des administrations*) moved to get the plan changed. In a report it attacked the dualities of several of the ministries and said that the 'circuit of decisions' from the Prime Minister down was too long. One of the cases it offered in evidence was the abortive battle the Finisterre authorities had had in October 1976 with the small German tanker *Böehlen* which they tried vainly and expensively to pump out.

GICAMA got some sympathetic action. That May the Cabinet decided to de-centralise POLMAR and give the *Préfet Maritime* and the prefect of the department powers to launch the plan on their own authority, and to work out their own locally co-ordinated schemes based on the protection of the most sensitive parts of the coastline. To give effect to this POLMAR was effectively divided in two: POLMAR-Mer to be run by the *Préfet Maritime* at sea; and POLMAR-Terre, also known as Plan ORSEC Pollution, by the prefect of the department on shore.

The funding of the plan was to be aided by the creation of a special intervention fund of £1,200,000 under the control of the Ministry of Culture and Environment, for emergency expenses. And the tender system could be by-passed by the use of a *certificat administratif*, in effect a simple contract to buy signed by a responsible civil servant. But it was not a perfect solution. By November as the Minister of the Interior, Christian Bonnet, was outlining the scheme in the National Assembly, there were still serious doubts about the efficacy of POLMAR among the deputies. The chain of command was still too long for decisions to be made quickly, the Assembly was told, and there was too much emphasis on 'consultation' and 'co-ordination' without a word about command. Indeed, no one appeared to be in overall authority when POLMAR was launched. What was particularly worrying was that the scheme had never been tested. Not only had there not been a real emergency, save the *Böehlen*, which in any case was strictly localised, but there had never been any realistic exercises. Some individual roles had been practised, such as the setting up of barrages and the use of 'skimmers' (mechanical devices which could separate oil from water). In 1975 an exercise had been held in Brittany assuming that a tanker had collided with a passenger ship off the Pointe du Raz and that south Finisterre was threatened with oil; but this had anticipated a spill of only

15,000 tonnes. A larger experiment, MINIPOL, off Marseilles, which was intended to use a controlled spill of real oil to test materials and treatment methods, had to be called off after protests from local politicians: it had been planned for spring just as the holiday season was about to open.

When the National Assembly commission of inquiry tried to find out what happened on these exercises later, it got what it called 'divergent and contradictory information'. Only two months after the ministerial instruction bringing the new POLMAR plan into being was issued on 29 November 1977, and only a few weeks before the *Amoco Cadiz* ran aground, a government working party report concluded that POLMAR still needed more attention to detail, particularly in the provision of equipment and anti-pollution products, if it were ever to work satisfactorily on the ground.

Even as the helicopters were still clattering out to the grounded tanker to take off her crew, the telephones were ringing in and around Paris to call the first meeting of CICOPH together. It began at 01.45 at Levallois-Perret where the sixteen-strong staff of CODISC and twelve telex operators, with their lines already chattering, waited to act as a communications centre. The meeting was chaired by the director of civil defence, Christian Gérondeau and his operations chief, Jean-François Di Chiara. Simultaneously in the Finisterre prefecture at Quimper, a similar meeting of regional officials was opening. This one was brought together by the Prefect of the Department, François Bourgin, but was chaired by Marc Becam, the slim and balding former agricultural trade unionist who was now mayor of Quimper but also a Minister of State at the Ministry of the Interior. Becam had been telephoned by his Minister, Christian Bonnet, and asked to meet with local POLMAR leaders to see what needed to be done. And he had arrived at the prefecture to find a POLMAR headquarters already being hastily put together from an office which had been equipped with telephone and telex links to gather results in the second round of the General Election due in two days time. Both meetings rapidly came to the same conclusion, that they had an unparalleled disaster on their hands. Already the initial damage to the tanker meant that they were looking at a spill more than twice as big as POLMAR was designed to handle. There were two priorities: to stop more oil coming out of the wreck and to protect the more sensitive parts of the coastline. Brittany, as the marine biologists hurried to remind them, was one of the richest seafood producing regions in Europe. After that there was going to have to be a massive clean up, not only to protect marine life in an area where the diversity of species was one of the greatest in the temperate world, but to protect the beaches on which the vital tourist industry depended.

There were only two ways to prevent more oil spilling: either to remove it or destroy it. CICOPH immediately voted against burning the wreck as a dangerous and politically inexpedient first measure when, as the British had seen from their attempts with the *Torrey Canyon*, it might not succeed. But removing the oil was a strong possibility: the committee was told that just after midnight Shell France had called CROSSMA at Jobourg suggesting the use of pumps to lighten the supertanker into smaller tankers lying nearby; and to that end the company had diverted the *Darina*, 65,000 tons; the *Niso*, 100,000 tons; and the *Halia*, 18,000 tons; all of whom were within a few hours sailing time. The first ship could be on station by 08.00 that morning and the others by the evening. At the same time a second call from Amoco had said it had located four powerful diesel pumps in Detroit which it was having flown to France. The Dutch company Smit International had been hired as salvors and two tugs were on their way to Brest to carry equipment for such an operation. The committee noted that two other other conditions were required, neither of which they yet had: time and calm weather.

Protection for the coastline meant floating barrages and a check of the national store of equipment and products for a POLMAR emergency printed as an appendix to the plan. It cannot have made very edifying reading. According to figures provided for the subsequent Senate commission of inquiry, POLMAR was intended to protect only 18.5 miles of coastline and that protection meant distributing the means so as to cover three different seaboards. Thus not only was the list of means slim, a great deal of it was hours, even days away from Brittany. On 17 March, it seemed that in the whole of France there were just eleven miles of floating barrage of three types, none compatible with the other, stored in eight different ports from Dunkirk to Marseilles. Of the total, just over half a mile of an air-inflatable type known as *Sycores* was at Brest. Other weapons were also thin on the ground: there were eight large skimmers of two types, the Vortex which remained stationary and the Cyclonet which was attached to a moving boat, of which only two, one Vortex and a Cyclonet attached to the support boat *Le Chamois*, were at Brest. The other six were distributed between Cherbourg and Marseilles. Each would be capable of removing about ten tonnes of oil per hour in calm weather. There were 1,000 tonnes of dispersants somewhere along the Channel coast; another 1,100 tonnes on the Atlantic and Mediterranean coasts; and 320 tonnes of chalk to use as a precipitant in Brittany itself. Of the dispersants, a dramatically controversial means of treatment in France, only 450 tonnes was in Brittany, say enough to treat 4,500 tonnes of oil. Thus it was clear to national and local officials that Brittany was going to need help.

CICOPH made three recommendations for action: first, that the pumping operation should begin as soon as possible; second, that every available foot of barrage should be got to Brittany as quickly as possible; third, that France should appeal for help to her neighbours, especially those who had signed the 1969 Bonn Accord, a mutual aid pact on pollution which included Britain, Belgium, Germany, Denmark, Norway and Sweden. Britain in particular would have to be told: not only did the two countries have an agreement to come to each other's aid in case of a pollution emergency in the English Channel, code-named MANCHEPLAN, but the predominant currents in those waters might well take some of the oil across to England in a mirror image of what had happened in the case of the *Torrey Canyon*. At any rate some of the oil might move towards the Channel Islands and the Navy operations room in either Cherbourg or Brest, PREMAR I and PREMAR II under the plan, would have to make a warning call to the Marine Emergencies Information Room of the Department of Trade in London.

The committee then began its secondary, but more important role: that of trying to put together the means that the ill-equipped Bretons would need on the shoreline. It was a time-consuming and patience-trying task that would seem to go on endlessly. And if there was a usable plan no one seemed to know about it. As Serge Chauvel-Leroux and Jean-Pierre Créssard reported in *Le Figaro*:

> For several days they found that they had not arranged any plan of action modulated according to the different qualities of oil likely to cause a *marée noire*: what products to use, what pumps, where to find them? In order to bring them together they had to have recourse to the most archaic weapon, the telephone directory, where they looked up, one by one, the numbers of all the suppliers of pumps in France and neighbouring countries. The same reaction came for booms, solvents, gloves and boots.

Since the future of the wreck was in the hands of the *Préfet Maritime*, Vice-Admiral Jacques Coulondres and POLMAR-Mer at Brest, the Quimper meeting was only interested in it in so far as it affected the amount of oil which might reach the beaches. Then Bourgin and the others began to piece together their own plans. It was not an easy task. Marc Becam recalled later that there had been 'numerous critics, but partly justified' and added that as regarded operations under the new POLMAR-Terre 'to my knowledge there did not even exist any documents at the municipal level, it is regrettable but it is so'. Nevertheless some tactics emerged from the discussion. To begin with Bourgin would keep his headquarters at Quimper, even though it was fifty miles from the wreck and thirty miles from Coulondres in Brest. So a mobile command vehicle from the defence base at Rennes would be summoned up and set up as an advance post next to the Gendarmerie station at Ploudalmézeau, just inland from Portsall, under the control of the under-prefect of Morlaix, Michel Saint-Prix. The under-prefects of Brest and Chateaulin would go to Brest as liaison officers with the Navy. The front line of the coastal operation would be four sectors, that were based on the towns of St Renan, Lannilis, Roscoff and Plougasnou, and co-ordinated by public works engineers from the Ministry of Equipment. The first priority would be the closing off of the oyster and shellfish beds, so as much barrage as could be found in the shortest possible time would have to be taken up to close off the Aber Ildut, Aber Benoit, and l'Aber Wrach, and then the bay of Morlaix.

After that it would be a question of gathering the men and materials to clean up the oil. Exactly what would be needed would depend on how far and which way the oil spread and on whether the Navy had any chance to do anything about the remaining oil in the wreck. Now in the small hours of the morning it was difficult to make any predictions, apart from one thing: with a spill of upwards of 80,000 tonnes they were going to need everything POLMAR could offer and a great deal more. The priority would be to find pumps of all kinds, pipeline, road tankers, trucks, and a massive quantity of simple materials like protective clothing, spades, rakes, buckets and other containers. The different administrations in harmony or conflict would be involved in bringing all these together; in preparing accommodation for advance teams; preparing for reinforcements of troops and civilians; checking on measures for collecting oil and evacuating it; and a dozen other tasks, in bad weather and in the politically uncertain conditions of the election interregnum. All this would take time.

It was against this background that a tactical decision was reached which, though it had some kind of elementary sense to it, was to cause uproar among the people and the media when it was made public. The spring equinox was approaching and each day on the coast the tides would reach higher and higher until, in ten days time, on Easter Sunday, 26 March, would come the highest flood tide. Until after that, it was decided, it would not be worth beginning a massive cleaning effort on the shores, simply because each day's work would be annihilated the next as the water carried the oil further inshore. The delay would enable a properly mounted effort to be put together. To the officials it seemed sensible, but to the public it seemed like incompetence or worse. And even in the two official Government inquiry reports later it was castigated as 'irresolution'. The local newspaper, *Ouest France*, was to suggest that the explanation was a cover for a paralysis of official will. The bluff was, it was suggested, that Plan POLMAR existed other than simply as a piece of paper and that officials were not improvising as they went along.

As the day went on in Finisterre more moves were being made. Michel d'Ornano, Minister of Culture and the Environment, touched down at Guipavas airport, outside Brest, in his Mystère 20 executive jet. He was taken by Becam and Bourgin on a flight over the wreck and then went to Navy headquarters where he announced the opening of the £1,200,000 intervention fund to pay for some of the costs of fighting the oil. Though the fund was a recent provision of POLMAR it was to be made up of credits from the four essential Ministries involved in its activities, Defence, Transport, Equipment, and Interior. What d'Ornano did not say in his announcement was that up to that day only £375,000 had been forthcoming from the other departments.

D'Ornano went on to join Becam and Coulondres for the most crucial of the day's meetings: that which would take a close look at what to do with the *Amoco Cadiz*. Navy planes and helicopters had been up over the wreck, spotting the dimensions of the slick and its condition, and boats had been scouting the possibility of starting the pumping operation. This was to be the first of a line of twice daily briefings which Coulondres would seek with his fellow officers, officials and advisers, in an almost exact duplication of the representation at the daily CICOPH meetings in Paris. At 15.00 hours d'Ornano and the others began looking at the alternatives before them, beginning once more with burning the wreck. There were precedents: in 1967 Royal Navy jets had strafed the hulk of the *Torrey Canyon* in exactly such an exercise, but the oil refused to burn, it had mixed too well with the water before the attack came. In 1970 the authorities had fired an oil exploration platform just off the Louisiana coast which had been spilling oil. As a result the platform was completely destroyed in a blaze which lasted 136 days. Whether it had saved any ecological impact was impossible to establish on a coast already well known for raised oil levels in the sea. In South Africa, fire had been used with success to eliminate 2,000 tonnes of crude oil following the grounding of the *Oriental Pioneer* in 1974, and 500 tonnes spilled by the Soviet cargo ship *Pati* which had hit a reef.

But now the experts in Brest began to line up against burning. The oil industry representatives, notably the IFP, estimated that even if it could be burnt, it would only destroy the thirty to forty per cent of the cargo which was most volatile, and these products, representing perhaps 80,000 tonnes of oil, were likely to evaporate into the atmosphere spontaneously in any case. The fire would also risk, through the tremendous black pall it would cause, a substantial pollution of the land as well, as droplets of tar sprayed down on to the fields of artichokes, cauliflowers, and potatoes and the whitewashed houses. There was also the remote chance that the whole wreck might explode; moreover, the oil slick had already reached villages on the coast, like Portsall, and there was a risk of a blaze shooting across the oil to engulf them. A huge black mushroom cloud, hanging for days and perhaps months over Finisterre would also be an extremely visible political embarrassment for the Government. Becam made the decision to rule out burning the oil.

The meeting then turned to the problem of pumping out the wreck. This was a practical proposition if they could only get a change in the weather. Time was of the essence because the oil was spilling and spreading very rapidly. Some of the facilities were already to hand in the form of the three Shell tankers. The four big pumps were being made ready for shipment from Detroit, the IFP was hunting France for floating pipeline, and the first of the tugs, the *Smit Houston* would arrive in the area the next day. But there was one serious drawback. The operation would take a minimum of five days once the pumps were aboard the *Amoco Cadiz*, and to pump quickly the small tankers would have to get as close in as possible to the wreck. But it appeared that the only local chart the Navy had was made in 1927 and its detail, for this kind of work, was sparse, having been drawn from soundings on a hundred metre grid. Even the rock the *Amoco Cadiz* had struck was not shown. No tanker master would take his vessel close into shore on the basis of such a chart. What was really needed was a new hydrographic survey of the seaward side of the wreck and then for an approach lane to be marked by a buoy-laying boat. Anticipating this, the Navy had sent out the hydrographic service ship *Astrolabe* and the Lighthouse Service's buoy-laying boat *Georges et Joly* during the early hours, but neither was able to get anywhere near the wreck to work because of the weather conditions. They would try again but the lack of the survey was a serious impediment to the operation. 'What we need is side-scan sonar,' sighed Coulondres.

Meanwhile there was the important job of fighting the oil that was already spilling into the sea. The light crude from the *Amoco Cadiz* seemed to have a propensity for spreading very rapidly and it was also being pushed quickly across the water by very strong winds. In theory oil was carried on water at about three per cent of wind speed, but here theory did not help: it had been almost impossible in the first daylight hours to predict which way the oil would go and how quickly. The Navy's first line of reaction had been to get a small flotilla of six boats into the area with instructions to dump chalk on to the oil to sink it, and to try spraying chemical dispersants in the hope of breaking it up before it could travel too far. The marine biologists at the meeting, including Lucien Laubier who was director of the Brittany Oceanological Centre, part of the CNEXO organisation, which had an impressive research centre on the coast near Le Conquet, were extremely worried about the use of dispersants and said so. The dispersants worked best on fresh oil. The more viscous it became as it weathered, the more likely it was that the spraying would have to fall back on an older, stronger but more toxic generation of chemicals. And there was the danger that unless they were used in the right quantities and under supervised conditions, the dispersed oil might simply rise again from the depths and recombine to form a new slick.

The Navy were nonplussed. Without dispersants how else were they to fight the oil? It would be almost impossible in the weather conditions off the coast to spread large quantities of such things as absorbents and agglomerates, to soak up the oil, and then collect them in; or even precipitants like the chalk. Their boats on the scene were already in difficulties. Skimmers and other mechanical oil recovery devices were almost non-existent and were notorious for being inoperable in waves over six feet high: out off the Men Goulven the seas were reaching twelve and fifteen feet that day.

A compromise was reached. The main danger was the use of dispersants close inshore where the highest proportion of marine life existed. If the boats stayed out of this area, if they did not spray in water any shallower than 150 feet, there would be a good chance that both dispersants and dispersed oil might be kept out of the critical zones. The limit would effectively keep the boats three miles offshore. The decision would also have to be passed on to POLMAR-Terre to stop the use of dispersants ashore; there must be nothing like the horrendous scenes in England during 1967 in which drums of dispersants were simply opened on cliff tops and poured over on to the rocks and beaches below. This strict limit on the use of dispersants was to have far-reaching consequences for the *Amoco Cadiz* operation.

The first of the Brittany people to learn about the wreck of the *Amoco Cadiz* were, ironically enough, not just those living closest to the scene. About eighty miles away in the port of Lorient on the south side of the Brittany peninsula in the early hours of the morning, police cars with loudspeakers toured the streets telling the public what had happened. They had to, for by the same freak of weather that had blown the smell of the oil down the coast to Lucien Laubier's house, the stench was hanging so strongly over the town that calls had been made to police headquarters asking what was wrong. For those who lived in and around Portsall, the *Amoco Cadiz* was a terrifying reality, squatting on the Men Goulven rocks only a mile or so offshore like some monstrous stranded shark, with an ominous pool of black bile spreading from its carcase which slopped into the little harbour in slow, hump-backed waves. The local people, the fishermen in the lead, ran and walked down to the breakwater and stared incredulously at the sea. Alphonse Arzel, mayor of Portsall and of Ploudalmézeau, the small market town just inland, was among them. He got back into his car, his eyes red and stinging with the fumes blown off the oil-laden water, and drove slowly back to his grey-painted office in the centre of Ploudalmézeau to tell his colleagues. 'I remember the shock, telling them the coast was black again was like breaking the news of the death of someone they loved,' he recalled later, 'it was as if the school bus had been in a dreadful accident that morning.'

As the morning got older, hundreds of people from the villages around came to the dunes at Tréompan, next to Portsall, to look out at the wreck. Most of them stood silently for a while, with the wind tearing at their hair, and then went home. But some, especially those who lived closest to the shore, went down on to the beaches and began a symbolic if rather pathetic attempt to help. Using garden spades, trowels, and even empty seashells, they began trying to scoop the oil into buckets and old paint tins. They were few and far between and they were on their own, and after a while, discouraged and angry, they left the sea to the oil again.

The reaction from most people in Brittany to the news that day was the same: shock followed by anger. Not only did they have the past experience to make them bitter about oil on the coast, but public opinion had been deliberately sensitised to the subject of possible threats to the environment during the bitter election campaigning of the past few weeks. What had been forecast had come to pass in a massive way. The anger was directed at the Government for not having done more to protect France, and more directly at the oil companies who were held directly responsible for the catastrophe. But the anger was also brought on by their own helplessness in the face of the spilling oil. The fishermen, hauling their boats as far inshore as they could get them to avoid the oil, had been kept ashore by the weather: but they knew there would be little point in going to sea in local waters; even if they could avoid oiling their gear, fish would be scarce for some time to come. On the quayside at Portsall they appeared to have little illusion about the abilities of the authorities and Plan POLMAR to tackle the oil. 'Here we go again, we will have the soldiers back with their buckets and their little spoons,' said one. Their union, the CFDT, reported, 'We put in bills for the damage caused by the *Olympic Bravery* and the *Böehlen*. But so far no one has received a penny.' They sent a telegram to President Valery Giscard d'Estaing reminding him that on his last visit to Brittany he had declared that, 'The coast must be protected. These events are a sort of dishonour to our civilisation and our first anxiety is to reduce the risks.' It seemed, the fishermen suggested, that all the steps the State had taken had been unable to prevent the same thing happening again.

The seaweed collectors saw their livelihood threatened, and they wondered when their harvesting season, due to start on 15 April, would begin this year, or if it would begin at all. 'Eleven years exactly after the *Torrey Canyon* disaster, what had been done to prevent the pollution from tankers? Nothing,' they complained. The oyster growers of the *abers* who had fought long and hard to protect their beds from a parasite infection that had hit other areas of Brittany now saw themselves prey to an invader of a different kind. And the hotel and guest house owners who had already been complaining about the late start to the season caused by the wintry weather, listened to the radio news bulletins and steeled themselves for the first telephone cancellations. More than 300,000 people depended directly or indirectly on employment generated by this most westerly coast of France.

But there was another mood behind the anger: a kind of solidarity in the face of adversity. It manifested itself first among the mayors of the communities along the coast. They too had been sizing up the extent of the disaster and had come to the conclusion already reached by the POLMAR leaders, that every available hand would be needed to fight the oil. So as the day wore on they waited for the regional administrators and government departments to contact them, to tell them what was happening and to ask for help. Their manpower and equipment were sparse and ill-adapted to the fight but they would do what they could, and their knowledge of the coast was unrivalled. Yet the calls did not come. More to the point, a check along the coast showed that very little seemed to be being done about the oil. In Portsall and to the south the fire brigades from Brest and district were trying to get on shore with their pumps and some small skimmers they used for road tanker accidents; sanitation wagons normally used for emptying cess-pits and street cleaning had been pressed into service as pumps; and a handful of road tankers had been brought down to the Portsall harbour side. But they could work only on the hard surfaces of the quaysides and so only while the sea was in the harbour itself, where more than a hundred tons of oil lay floating on the surface. There was feverish activity at the seaward ends of the *abers*, where floating barrages were being hauled into place: over half a mile across the Aber Benoit and just less across l'Aber Wrach. But that apart the beaches and coves seemed deserted.

In the meantime they could see local people doing what they could to help themselves: in the oyster beds, the growers who had facilities were making frantic efforts to ship their cages of oysters to new beds in the south of Brittany; firms who kept lobsters in keep tanks for shipment to restaurants were doing the same. And so they began trying to bring together their own teams of workers, recruiting them from the fishermen and the farmers. The tankers at Portsall were soon filled with the melange of oil and seawater from the harbour and then had to make the long drive to Brest where a disposal area had been allocated for collected material in the port area. Each tanker was lucky if it made two round trips in the day. In the meantime, the workmen, sometimes up to their thighs in the black water and hit by stinging eyes, dizziness, headaches and nausea from the fumes, cleared an area of weed for the next pump. As the tides began to change the oil was going further and further out of reach of the pumps and was becoming too thinly spread on the water to be pumped successfully. But at least, they felt, something was being done.

It did nothing, however, to assuage the anger of the mayors about not being consulted. By early evening, a meeting of a representative group, mayors of all the communities from Ploudalmézeau to Lannilis, had come together with Alphonse Arzel in the chair to make their presence felt. They sent a strong message to Prefect Bourgin, asking for representation for the mayors at the POLMAR meetings. They also agreed to set up a fund to help the little village that was bearing the brunt of the oil: they called it 'SOS Pollution Portsall'.

But even while the meeting was going on, the anger of the local people was beginning to spill over, too. Two thousand of them converged on the blackened harbour of Portsall to demonstrate against the *marée noire*. Small children carried hastily scrawled placards: 'What kind of Brittany will I have in twenty years?', while their parents had cards which demanded simply 'Halt pollution'. Local union leaders and politicians, their banners waving above the crowd and ever mindful of the election only two days away, began to get a word in.

The local fishermen, drinking to their sorrows in the local café, *Le Récif* (pointedly its name translates as The Reef), watched the evening television news as Michel d'Ornano's explanation of POLMAR tactics was reported and shook their heads. In particular the mention of dispersants brought calls of disagreement: no one had yet explained the evolution of such products to them. They simply remembered 1967. But there was one thing they did know about, they said. Pumping out the tanker would never work. They knew the local waters too well. One man jerked his thumb in the direction of the *Amoco Cadiz*. 'If the sea has made up its mind it wants her, it will want all of her.' Jules Légendre, deputy mayor of the village, who had sailed for twenty-three years in tankers, gave his view: 'The best thing would be if they bombed the wreck now.'

By mid-morning on Saturday 18 March the slick had almost trebled in size. It stretched in one continuous sheet 18½ miles south to just beyond Le Conquet and then north-eastward across the mouths of the barrage-protected Aber Benoit and l'Aber Wrach to the Ile de Vierge, 7½ miles to the east. The thickest part of the slick was around the wreck and plastered against the coast at Portsall. But something else had happened: the wind had backed to the south west and was beginning to move the oil north again and out to sea. Already it was ten miles from shore and still moving north and east. But the wind speed had dropped temporarily to only ten to fifteen knots and there was a weak blue sky over a restlessly stirring ocean. The stench of the oil was still powerful. People living in Quimper and Lorient still claimed they could smell it during the night and the pilots of planes passing over at 6,000 feet reported it strong even at that height.

The Navy's morning briefing in the sparse, vaulted office in the underground labyrinth of the prefecture, which had hastily been equipped with trestle tables and charts as a POLMAR-Mer control centre, was a bleak affair. The previous day's spraying and dumping of precipitant had made no observable impact on the oil slick. What was more the boats had been operating almost blind, able to tackle only the oil within sight from their bridges without knowing which way the thickest deposits were moving. Helicopters would be needed to overfly the area from now on, to act as spotters and try to predict the movement of the oil.

And there would be more help in this respect; CNEXO, IFP and the National Geographic Institute (ING) would be making regular flights over the slick in a B17 taking thousands of infra-red pictures. But more boats to tackle the oil would be needed. The British might provide some under MANCHEPLAN but they would have to find most of them themselves. The lull in the bad weather was tantalising: if they had had all the equipment together for the pumping operation and a good chart it was just possible they might have made a start. Six tonnes of floating pipeline had come in overnight from Amsterdam, but there was no sign of the pumps from Detroit or any chance of getting them more quickly from any alternative source. Nor did they have their survey of the reef. At first light a team of frogmen had gone out to the wreck in an outboard-powered Zodiac dinghy to try to find out more about the damage to the hull. All they could do was confirm what was already known: that the ship had torn open at the bulkhead between the number three and number four tanks. But they managed some useful work, making a rough plot of the position of the rocks for almost one hundred yards around the vessel and taking depth soundings.

On land it still appeared to the local people that there was no organised oil fighting operation. As the small contingents of sanitation wagons went back down to the water's edge, the fishermen called out to the drivers, 'Where's the Government?' The drivers only shrugged. The local mayors had been talking again and were clearly contemptuous of what had been done the first day and of the authorities' tactics. If they were going to try to collect the oil as a first line of defence, much more equipment would be needed. And if POLMAR did not know where to find it, they had an idea of their own. There were other weapons to hand in a farming area like Finisterre. Pig breeding was one of the biggest single farming activities along the coast and almost every farmer had what the French called a *tonne à lisier* and the Americans a 'honeywagon', used to pump out cesspools and clear waste from the sties. Each could hold between 500 and 1,000 gallons and was easily manoeuvrable, being pulled behind a tractor. Honeywagons and tractors could travel along the beach surface and follow the oil as it was being carried by the tide. Though the mayors, as they talked on the phone to the local farmers' syndicates, may not have fully realised it at that moment, the use of the wagons would revolutionise the operation, for not only were there hundreds to call upon, they had the vital additional ability to work during a greater portion of the tidal cycle than the road vehicles, which were restricted to hard surfaces. Introduced by the men POLMAR had not thought it worthwhile talking to, the honeywagons were to become perhaps the most important single mechanical weapon on the entire *Amoco Cadiz* operation.

Certainly POLMAR was having its problems. As the hours went by equipment was beginning to be found, but co-ordination was poor. The headquarters at Quimper was being overwhelmed with competing offers of assistance, some sensible and practical but sometimes mutually exclusive, and others quite absurd. There was the man, for instance, who suggested using two aircraft carriers, joined by a massive length of floating boom, which would surround the oil and tow it out into the Atlantic, where planes from the flight decks could bombard the oil with chalk and sink it. Private companies rang from all over the world to offer their professional services: one American firm offered to clear the entire slick for £25,000 a day, while a Dutch company said it could do the job for a round sum of £2.8 million. At the beachhead equipment was not always where it was supposed to be. A great deal of it seemed to be getting stockpiled on Quay No 1 at Brest docks; pumps arrived where the oil was too thin to be pumped; skimmers where there was pounding surf. But practical help came too; more lengths of floating barrage, and some specialist manpower in the form of seventy men from a special civil defence unit, known as UISC7, based at Brignoles in the Var. Created in 1974, UISC7 was a unit of the Army engineers put at the disposal of the Ministry of the Interior as a civil defence commando force. The 462 man regiment was trained to teach civil defence techniques, but also operated as a crack mobile unit, being flown in to areas ravaged by forest fires or into the Alps in the avalanche season. Though it had no direct pollution fighting expertise, its organisational and leadership abilities would be important. A major hold up in the organisation was that the mobile advance post at Ploudalmézeau was still not in operation, and all communications were having to be handled through Quimper. The result was that officials of different administrations made their own decisions and sought solutions to any problems they faced at local level, sometimes to the detriment of each other.

The *Amoco Cadiz* had become the subject of world-wide attention and Guipavas, a single runway with a small terminal building, suddenly became the landing place for an air force of chartered planes as the world's press and television, oil company executives, pollution experts, scientists, lawyers looking hungrily for clients, salesmen and others flew in. Hotels in the centre of Brest, normally quiet, suddenly found themselves filled to bursting, their lobbies packed with shoving throngs of men, luggage and television equipment. The first news reports from the scene, shouted over the hotel phones, gave the impression that all Brittany was black from end to end with oil. The tenor of these reports would become important. Meanwhile in the streets outside, two thousand trades unionists and conservationists marched in protest to the Marine Prefecture. The bodies of oiled seabirds, collected from the beaches, were waved angrily at the Navy's door.

But the greatest anger seemed to come from the fishermen. Although only one of the groups whose livelihood might be affected by the oil, their mood was the more noticeable for their reputation for stolid phlegmatism. Already four hundred of them had appeared, unusually, among the two thousand who had protested in Portsall the afternoon before. Now their leaders, headed by the fiery, moustachioed Henry Didou, secretary of the Brest Fishermen's Committee, sought out Jean Chapon, secretary general of the Merchant Marine, the government department which controlled their activities, who had arrived in Brest and told him they were ready to hand in their licences in protest at government inactivity. The threat was a serious one, for not only was the licence, known as a *rôle d'équipage*, a permit to work, it was also a social security card. They also sought, and got, a promise from the startled Chapon that the Government would look urgently at the setting up of an indemnity scheme to help them with their lost income. It would be needed, explained Didou. The previous day a fisherman at Portsall who had just mortgaged his entire future to buy a new boat had tried to commit suicide.

Chapon was only one of a number of central government officials and politicians who had been flying into Brittany anxious to be seen to be doing something. And in the middle of the day, ever mindful of the looming election, came Prime Minister Raymond Barre himself. Barre, who had been a professor of economics and was a strong admirer of de Gaulle, was a former Transport Minister who had earned a reputation in government for never courting popularity and never shying away from telling the people the truth, however unpalatable it might be. He had insisted for months after joining the Government that he was not really a politician and the French people had warmed to that sort of approach. Now it seemed he was gambling that his visit, reinforcing solidarity with the local people in a traditionally strong Gaullist area in time of trouble, would be seen as another example of his popular pragmatism. It was arranged that he should meet with local leaders and the town council at the little town of Landéda between Aber Benoit and l'Aber Wrach. But the meeting did not go at all well.

In the beginning it was so impromptu that arrangements were still being made to receive him while he was still in the air. Even as he was arriving a hunt was going on in the town hall for a picture of President Giscard d'Estaing to hang on the wall. It was found behind a cupboard and hurriedly put in place. Barre touched down at Guipavas, accompanied by Interior Minister Christian Bonnet, to be met by Marc Becam and François Bourgin. The four were taken by helicopter to a sports field in Landéda and then transferred to cars to be taken to the town hall. They drove up to meet an angry crowd of two hundred demonstrators on the front steps, waving the black and white Breton flag. So hostile was the reception, with Barre actually being hit with an umbrella as he tried to announce that, 'Everything will be done to limit the consequences of this pollution,' that the mayor, Alfred Marec, felt it necessary to sit down the next day and write a personal letter of apology. Inside, the official reception was no warmer: an angry town council began to attack him on the lack of facilities available to tackle the oil, and demanded that ships of flags of convenience should be banned from territorial waters. Dr Yan Daumer, an assistant to Francis Le Blé, the Socialist mayor of Brest, attacked Barre bitterly for the Government's lack of provision for preventing such a catastrophe, claiming sarcastically, 'It's always apples for the pigs, but peelings for the Bretons.' But an increasingly angry Barre hit back at this, arguing, somewhat hollowly, against 'the use of this drama at sea for electoral and political purposes'. Daumer reminded him sharply that most of the likely successful candidates along this coast would be from the Government majority coalition.

When Barre found himself under attack from Louis Le Roux, the Communist deputy mayor of Brest, he abruptly called his visit to an end and left the room, refusing to speak to a deputation of oyster growers' representatives from the *abers*. Barre sat fuming in Marec's office while gendarmes scurried to find his car. Outside the town hall the fierce argument was still going on, this time between a local Socialist Deputy and a party of young Giscardians. Barre and Bonnet got back into their car, slammed the doors and left.

But Barre had made one important announcement. Neither he nor Bonnet were foolish enough to believe that the complaints the local politicians had made about POLMAR did not have the ring of truth. The co-ordination between the national and local efforts would therefore be tightened. Christian Gérondeau, the director of civil defence, and his deputy Jean-François Di Chiara, who had been operating from CODISC with CICOPH, would be moved down to Brittany. They would move into the new advance post which was to be set up at Ploudalmézeau.

Behind all this, the two main protagonists in the *Amoco Cadiz* affair had not escaped attention. When the *Pacific* finally put into Brest at 15.30 hours local time, Weinert found a *juge d'instruction*, Jean-André Gouyette, the West German consul and two court officials waiting for him. He was taken to the nearby gendarmerie maritime barracks, where Bardari was also being held. Gouyette had already talked to the fatigued and hollow-eyed Bardari and Weinert was now told that they were both being placed under open arrest to see if either or both of them should be charged under French law with negligently polluting the seas.

The maximum penalty for the offence was a heavy fine or even up to two years imprisonment. Nor was it the only procedure they would face. A naval inquiry had already been opened into the incident by the chief administrator of the Marine Affairs Department in Brest, Guillaume Fertil, in which Bardari and his fellow officers in particular would be deemed 'privileged witnesses'. Fertil had already been questioning Bardari for several hours and, to Gouyette's consternation, had ordered the papers of both the *Amoco Cadiz* and the *Pacific* seized. On the basis of Fertil's report, the Government would decide whether to take action in the maritime court for breach of the French Merchant Marine disciplinary code. To emphasise the strength of these actions, the *Pacific* would be impounded by the court. The development was transmitted back to Amoco and to Bugsier, who immediately ordered the *Simson* to leave French waters. Gouyette and his officials then settled down to an interrogation session which was to last until late into the night.

On Sunday 19 March, the French went to the polls and nationally the *Amoco Cadiz* affair temporarily shrank in significance compared with the issue of where the country was going. But the tanker disaster could hardly be avoided as an issue and there were angry accusations and counter accusations of parties using the 'sadness of Brittany' to make political capital.

It was the showing of the smaller parties, whose share of the vote might well decide who formed the next majority in the National Assembly, which remained the unknown factor and still fascinated the political commentators. Among them were the Ecologists. Their candidates had not done as well in the first round as they had during the municipal elections of 1977, but the question now was: given the possible extra boost to their cause from the *Amoco Cadiz*, would they be able to persuade voters to support them when national rather than local politics were involved. They had had some setbacks in the interregnum, especially when one of their best known leaders, Brice Lalonde, had inexplicably called on his followers to vote for the Union of the Left as the most practical route to ensuring the kind of policies the Ecologists wanted. In Brittany the question was: with the first round results close in many constituencies, would the *Amoco Cadiz* become such a predominant local issue that anger against the Government would let in the Left?

Even as the voters made their way to the polls, new moves were being made in the fight against the oil. More oil than ever now appeared to be spilling from the wreck: Charles Pavot, mayor of Porspoder went out in a small boat and reported 'at every wave you can see a small black geyser,' and the slick was moving remorselessly eastward under the effects of the current and a twelve to twenty knot west wind.

That morning Navy frogmen had twice gone out to the tanker to try to find out more about where she was holed, but without success. An advance team from Amoco International had flown in; led by Harry Rinkema, the company's tough vice-president for marine transportation, it also included his operations manager, Captain Claude Phillips, who had talked with Bardari aboard the *Amoco Cadiz* in the hours before she grounded; and a small posse of lawyers, public relations men, pollution and salvage experts. Coulondres had made a call to Amoco in the first hours after the wreck and made it quite clear that the company would have to come to help clear up its own mess in France, but the team was already being put together. 'Our job is to stop the flow and get the oil off,' Rinkema told reporters.

The manpower of POLMAR-Terre was being strengthened: two companies of soldiers from the Third Military Region at Rennes, totalling 280 men, were moving in, one company to be based at Ploudalmézeau and the other at Lannilis. The mobile command vehicle had finally been set up in a sports park at Ploudalmézeau. And there was more local equipment and volunteers: fifteen honeywagons were now working at Portsall and thirteen road tankers each capable of holding twenty tons of material had been found.

Compared with the amount of oil spilled it was still a pitiful presence. The tankers, filled from the honeywagons, found that they could only make two round trips each in the day to the disposal site at Brest docks, limiting the beach teams to just over five hundred tonnes moved. At sea, POLMAR-Mer still had not been able to get its survey of the waters off the reef, and there was a powerful swell running. The boats seemed to be working more effectively because of the aerial spotting of the oil and the Navy were now seeing if the use of the CNEXO infra-red photos, plus accurate weather forecasting and tide tables, would enable them to get a more accurate prediction of what would happen to the oil during the night. The Navy operations room in Cherbourg had been told of the oil moving east and out to sea and under MANCHEPLAN a warning call had been issued to the Channel Islands and London asked what aid it could give. Ironically, MANCHEPLAN too was in a state of flux. Only two weeks before, British and French officials had met at Cherbourg to put the finishing touches to the latest version of the plan, and the Brest operations room found it had not even seen a draft. But the British at least were ready to take action. They began to mobilise inshore craft along the south coast of England and to put dispersants aboard. And they began to order Royal Navy tugs to sea to work to a forward post which would be established at St Peter Port on Guernsey.

The progress of the oil had been watched with increasing alarm by the local authorities in the neighbouring department of the Côtes du Nord. They knew what an oil slick was about: it had been their Pink Granite Coast which had been covered by the 30,000 tonnes from the *Torrey Canyon* and they now feared they were in for something far worse. Already some of the holiday resorts which had important amenity beaches had taken unilateral action: they had hired scrapers and bulldozers to peel back the top few feet of the beach sand and pile it up above th high water mark. The sand would be kept there until the oil threat had disappeared and then laid back down. As the oil had moved eastward towards them they had seen the Finisterre authorities setting up forward posts at Roscoff, Plougasnou and Plouescat. On rough calculations if the oil spread at its present rate it would be with them within three days. The mayors of the coastal towns and villages pleaded with the department authorities to launch POLMAR in the Côtes du Nord. But no moves were made.

In the afternoon, while Rinkema was visiting Bardari, to get his first-hand report of what had happened aboard the *Amoco Cadiz*, the sad spectacle of the captain's former command took on a new role, that of a tourist attraction. It became, in the words of the local newspapers 'the star of the weekend'. With the intermittent driving rain of the past few days having lifted, and a brighter sky, thousands of people had made for the coast to look at the ship and the oil. So many turned out that at one stage in the afternoon there were traffic jams and parked cars on every road to the shore between Porspoder and Plougerneau. It took between one and two hours to get from one village to the other, only twelve miles apart, by car. A column of cars nearly thirteen miles long followed one after another through Plougerneau, Le Conquet and down to Brest.

All along the stretch of coast, the police fought to keep the traffic moving, and to make way for each road tanker full of oil from the beach which was taken, with some grandeur, under police motor-cycle escort, headlights blazing, off towards Brest. At Portsall itself, the police simply sealed off the village and turned cars away, so the crowds drove around the headland to the dunes of Tréompan or Lampaul. There they trekked across the sand, most on foot, but many of the teenagers on their revving motor-bikes in an impromptu motor-cross event, to get a view of the ship, for all the world as if they were on their way to a fair. 'It only lacked the smell of frying onions and sausages,' reported one newspaper. The Ploudalmézeau local council, fearing something like this, had in fact banned access to the dunes, but under the sheer weight of numbers there was no way it could be enforced. The best spots for a view were reached by knocking down the signs placed by those who lived nearby, who had turned out with their spades and trowels on the first morning: 'The sea is done for, preserve our dunes.' *Ouest France*, the regional morning newspaper, caught the atmosphere with a page top headline: 'The assault of the rabble'. What seemed to annoy the paper most was that the number plates on the cars showed that the majority of them were from Finisterre.

France awoke on Monday 20 March to find the Government still in power. A record eighty-five per cent turnout had delivered a decisive victory. The voters had decided to play safe rather than plunge into the unknown – a tendency so well divined by the wily de Gaulle when he built the week's delay for 'reflection' into the electoral system – and the balance between the two coalitions had come out remarkably unchanged. None of the seats seemed to have turned on the question of the *Amoco Cadiz*. Even with a major chance to express their opinions to their leaders in the clearest possible way only days after the disaster, the voters of west Finisterre in particular had gone along with the party line.

But those involved in POLMAR did not have long to mull over the results; the weather had deteriorated again with fifty knot winds from the south west and west and a surging ten foot swell, and they found themselves facing the first of a series of difficulties which would have important consequences: the floating boom on l'Aber Wrach had failed and the oil was being carried towards the precious oyster beds. It was to be followed in the next two days by those on the Aber Ildut and Aber Benoit and, later, others. They would be mended quickly but booms would remain a controversial subject throughout the operation. Following POLMAR strategy, as much floating barrage as could be found was being rushed by road to Brittany; another four miles had arrived that day; and as quickly as the cumbersome arrangement for handling it would allow, it was being laid along the coast in advance of the drifting oil. The importance of the *abers* having been realised at once – there were about 16,500 tonnes of oysters in the beds with a harvest between March and August of 1977 which had been worth over £280,000 – a team from Equipment and the Lighthouse Service had been battling to lay booms there on the first day, and all the biggest beds had been closed off before the oil arrived. But the boom laying operation as a whole was not going well. In the beginning, of the twelve miles or so of boom held in France for just such an emergency, only six had been in good enough condition to use. Most of the first boom to arrive was of the French-made *Sycores* type, which, air-filled and light, was too fragile for the heavy sea conditions in which it was being used; and the deep rubber skirt hung beneath it created exceptionally large drag forces which damaged the boom fabric and connectors.

A light aluminium type for use in harbours proved hardly effective at all; and even the best version, a foam-pellet-filled model called Acorn, was often ineffective because few of the men appeared to have been given any special training or instruction. Some booms were laid in areas of strong current in such a way that the oil was simply carried beneath their skirts by entrainment; deployment was often based on equity of protection rather than choosing the best boom for the job or the best way of laying it; many were laid too near rocks where they could be damaged, or where they could be fouled by debris, or where they were difficult to anchor (ironically, permanent seabed anchorage points for booms in the *abers* had been one of the long term notions for the local POLMAR plan). And quite often they went untended, the laying team simply having completed its work and moved on, so that oil accumulated on the seaward side and spilled through the booms into the waters they were supposed to protect. The secret, as British and Canadian experts who flew in later explained, was not to lay the boom straight across the inlet, but to angle it according to tide and currents; this would reduce the effective speed of the water flowing past and would also tend to trap oil at one end where it could be pumped away as it built up. Also many areas should have had several booms, one after the other, rather than just one. In other areas where the tidal range was large, the barrages should have been manned so they could be adjusted for each change or drifted with th current. But overall, the weather simply proved too much for most of the booms and the nature of the oil, light and easily dispersed in the water column, allowed it to pass the barrages.

POLMAR administrators would maintain later that they had never said the booms would stop the oil, only divert it so that it would drift further down the coast, but that was not an explanation understood by those whose livelihoods depended on the oil being kept out to sea. And even as the booms were failing in the *abers*, local oyster fishermen were helping in a battle to get them in place across the bay of Morlaix, to protect 35,000 tonnes of oysters, a harvest which had been worth nearly £700,000 in 1977, and other shellfish. The failure of the booms along the coast to do what was seen as their job, caused widespread anger and resentment, which continued even after the initial mistakes had been corrected and the Navy had been persuaded to lend twelve teams of marine commandos with Zodiac dinghies to tend them. In the coming days, oil would reach all the oyster beds, killing thousands or, more often, tainting them so as to make them inedible and compromising the entire future of the local oyster growing industry with the workers it employed. The trouble with the situation was, the National Assembly inquiry was to decide later, that everyone had somehow come to expect a great deal more from the booms than they were able to deliver, treating them, in its own words as 'a veritable floating Maginot Line'.

In Brest, Rinkema and Phillips had been talking again with the Navy about the operation to pump oil from the wreck. The four pumps had arrived from Detroit and were on board the tug *Smit Polzee*, and a second tug had arrived from Lisbon so they now had all the equipment they wanted to begin the attempt.

Five or six days would be needed to set up the apparatus at sea and there now appeared to be some discussion on tactics: whether the oil should be pumped out into Shell's 18,000 ton tanker now anchored in Lyme Bay, a few hours away on the English coast, or whether it should go into a series of smaller 6,000 ton vessels which might be easier to get in close to the reef. The discussion was academic for the moment. For, as Rinkema explained grimly after returning to his hotel, yet another attempt to get a survey of the waters around the wreck, by the *Astrolabe*, had been stopped by the vicious weather. 'We can't even get a helicopter up over her', he commented sourly, 'We have the equipment and we are ready and waiting'. Both he and Phillips had been to see Coulondres and 'we gave him a commitment to do everything possible to stop the oil coming out and to take the oil off as soon as we can.' What was worrying now was the amount of oil which might already have come out of the *Amoco Cadiz*. So far it was not known exactly where the tanker was holed, other than where her back was broken, and it was being assumed, said Rinkema, that four central tanks and five smaller tanks, including the bunker fuel tanks in the rear section had been breached. Rinkema did not do the simple arithmetic, but that damage suggested that more than 100,000 tons, or approaching fifty per cent of the cargo of the *Amoco Cadiz* was already in the water.

At a hastily convened press conference Rinkema and Phillips were questioned about the responsibility for what had happened at sea on 16 March. Phillips replied curtly: 'Amoco's policy is that the master has absolute responsibility for the safety of his ship. We see no reason to change that policy because of what has happened.'

The French seemed to agree. After forty-eight hours of close questioning, Pasquale Bardari was taken before judge Jean André Gouyette once more and formally charged with negligently polluting the seas. He was released on £21,000 bail. Weinert was not immediately charged, but Gouyette made it plain that he was not yet finished with him and he ordered the German captain not to leave Brest. Rinkema commented: 'We stand by our captain. He respected the rules and traditions of the Merchant Navy.'

By the afternoon of Tuesday 21 March, the oil reached nearly fifty miles eastward from Portsall and the first thin tongues went deep into the Bay of Lannion, tainting the Côtes du Nord for the first time and twenty-four

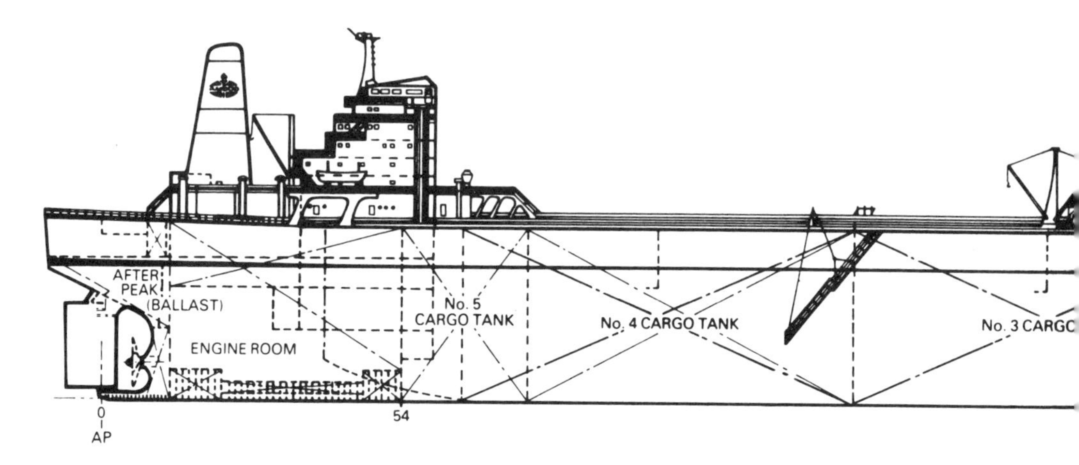

A side view of the *Amoco Cadiz* showing the division of the cargo tanks. The four larger tanks (1–4) are divided laterally into Port, Centre and Starboard tanks. The aftermost (No 5) has only two smaller Port and Starboard tanks flanking the pumproom.

hours ahead of prediction. The westerly winds and currents which ran parallel to the shore had carried the oil swiftly over the weekend, pulling it away from the coast to the south and west of the wreck and moving it east. The infra-red photos from CNEXO flights showed the heaviest deposits trailing away in ragged streamers from the wreck itself; in and around the port of Roscoff; around Plougasnou and on the Pointe de Locquirec in the Bay of Lannion itself. Huge sheets of oil of medium thickness extended close to the coast from Portsall to l'Aber Wrach; from Brignogan to Plouescat and in patches out to sea off Roscoff and the Ile de Batz, Plougasnou and Locquirec. And a thin sheen, far greater in area, extended from Porspoder, south of the wreck along the entire length of the polluted coast, pushing across the extensive sand flats in the Bay of Lannion towards the little towns of St Michel en Grève and St Efflam. As the scientists had predicted, the oil was being driven mostly on to the west facing shores.

In the face of the huge slick, hesitant POLMAR-Terre tactics had been concentrated for the moment on using the limited but slowly growing resources (six new mobile pumps had arrived and road tankers and barrages were coming in from Alsace, Provence, Belgium and Holland) to protect the creeks and inlets and only cleaning the most heavily polluted spots, known in operational slang as 'third degree' oiling.

Thus from Pointe de St Mathieu in the far south round to Argenton west

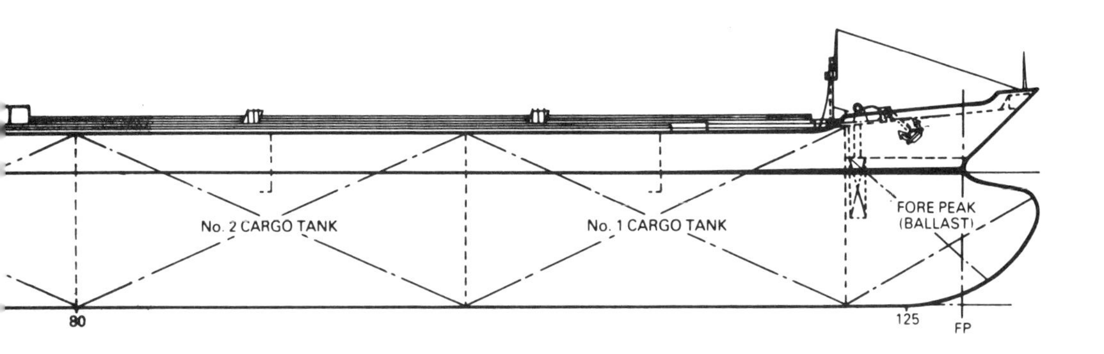

of Portsall the pollution was considered light enough not to clean. Portsall itself was the centre of the concentrated pumping effort, a melée of road tankers, tractors and trailers, pipeline and men in oilskins, all filthy from the oil, which was also now covering the roads down to the harbour and spraying up on the white-washed cottages. Eastward to the Ile de Sieck, near Roscoff, there was no cleaning; in Roscoff UISC7 men were leading a smaller pumping operation; and on the other side of Morlaix bay, another civil defence team from Bar le Duc were pumping at Plougasnou. In Morlaix bay itself, nearly two miles of floating barrage was being pulled into place, but more would be needed yet to add to the protection and an important battle would develop through the week to keep out the oil. Another half mile of barrage brought on site was found to be in too bad a condition to use. In Côtes du Nord, 150 firemen in three flying columns were being organised as the first line of defence and a small team of eighteen Belgian civil defence workers were helping with barrages to protect oyster beds and other areas in the Bay of Lannion. But there was still confusion: a company of soldiers who had only arrived to help with the work on the Sunday at midday from the 1st RIMA de Granville regiment were told to pack their bags; they were being replaced with men from the 41st de La Lande d'Ouee.

Nor did the organisation of the operation appear to be being dealt with decisively enough to suit the local mayors. After their angry meeting on the first day and their message to Bourgin, POLMAR leaders had kept them closely in touch with what was going on: there had already been one meeting at the prefecture in Quimper, and that day another, with Becam, at the Brest Chamber of Commerce. Yet their organisation of the honey-wagons apart, they still had not been given any specific role to play in

POLMAR, and they felt their views on tactics and strategy ought to be listened to. Now a strong section of them led by Charles Pavot of Porspoder and Jules Legendre of Portsall were calling for something very drastic: the deliberate opening up of the wreck and a full scale clean up immediately, with the fishermen and the farmers helping the troops. Startled, Becam and his colleagues turned the idea down. Legendre, red eyed from the oil fumes in the air round his little office in Portsall and from lack of sleep, said angrily: 'Sometimes I think I am banging my head against a wall. I was twenty-three years in tankers and I tried to tell them that all the oil would be out of the wreck anyway in a few days, at the weekend at the latest'. Legendre said he had asked Becam to get the Government to put out an international call to Europe for manpower to clean up the oil, but this did not go down well either, 'I was told we would look fools if the tanks held. If it happened, they said, we would always have time to launch an SOS!' Legendre leaned forward, 'All I want to do is to save my village and I will do that to the last ounce of my strength.'

At sea both the oil and the weather were posing serious problems for Coulondres and POLMAR-Mer. He had been able to increase his fleet of spraying boats from six to ten when four minesweepers arrived from Cherbourg and another six had been promised by the British under MANCHEPLAN, but their work would have to be held up while French laboratories checked out a new product, Dasic, that they wanted to use on the oil. Unlike Britain, France had no official system for homologation of dispersants and so each had to be examined before it could be passed for use. The main chemicals being sprayed, made by BP and Fina, were luckily already well known. Coulondres's biggest problem was that the thickest oil was closest to the shore, and that was where his boats could not go: first because of the uncharted rocks and secondly because of the 150 foot depth rule for the use of dispersants. Effectively his craft were often being kept three to four miles off shore.

The weather still prevented the tanker pumping operation. A team of frogmen had been out to the *Amoco Cadiz* during the morning and found an important new spill of oil; the rear portion of the wreck was being moved by fifteen-foot waves, it was apparently 'shaking' under their force, but the forward portion did not appear to have moved. Exactly what had happened they were unable to find out because the sea was too strong to allow them to leave their dinghies. They had some help in the shape of eight salvage men from Smit who were dropped on to the forward deck of the tanker by Navy helicopter.

The Dutchmen had been trying to make their survey for some time but the French would not give them the use of the helicopter. In exasperation one of the Smit men had taken out a boat and gone aboard by climbing up the ship's anchor chain. After that Smit were allowed two machines. From his own men's report and what Smit told him, Coulondres make a devastating assessment: as much as 150,000 tons of the oil might now already be in the sea. At least three tanks seemed to be holed on the starboard side of the forward section. Yet the official policy remained at the moment to pump out the wreck. The problem of getting the lightening tankers up to the wreck was still a major stumbling block. Coulondres still needed his hydrographic survey, but even the most sophisticated vessels would have found it impossible work in forty knot winds, he admitted. Neither were the tides any help. As they grew higher each day, they enabled the sea to cover more and more of the wreck. Even if the conditions were suddenly easy enough to allow pumping, the operation could not be continuous, the pumps would have to be lifted off every day by helicopter as the water closed in.

But it emerged that other minds had been applied to that problem. Quietly the French Petroleum Institute (IFP) and a team of local engineers, the same group which had worked together on the attempt to pump out the *Böehlen*, had put together an alternative scheme which they now laid before Coulondres. Instead of putting the American pumps aboard, they envisaged getting a power source out to the *Amoco Cadiz* and using her own pumps.

The oil would go down a pipeline either to a tanker well off the rocks or possibly even directly ashore. The engineers themselves admitted that there were still a number of unanswered technical questions, for instance whether the ship's pumps were powerful enough to overcome the friction caused by the oil passing along the pipeline, for up to two miles might be needed. But if it were agreed, they insisted, arrangements could be made within three days and, once they could get aboard, to empty the wreck might take eight. No one seemed to have queried the practicality of a scheme that relied on a pump room that had been underwater since the tanker had grounded, using pump lines inside the ship almost surely ruptured by the damage wrought by sea and rocks. But Coulondres agreed to let them try the scheme. It was as well to have all the weapons he could call on and, if it were pulled off, he would have the political advantage of a French operation more closely under his control than that of the American owners and Dutch salvors. 'Only God knows if we shall succeed,' said the Admiral.

But overnight there appeared to be an observable change of mood and by the morning of Wednesday 22 March there was an air of quiet desperation at Navy headquarters. With the weather unchanged, two frogmen had gone out to the wreck and found that, though some forward tanks still contained oil, the ship was now beginning to move, with the bow tilting upwards visibly. The reports, carefully analysed by the Amoco experts, seemed to suggest only one thing: that all the tanks of the ship must now be, directly or indirectly, open to the sea. More and more calls came from sources on shore expressing open contempt at what POLMAR-Mer was doing and demanding that the pumping operation be abandoned and new tactics tried. And the Navy found itself making concessions to the view.

They asked Rinkema and Amoco to 'consider and comment on controlled burning of the wreck'. The French themselves went into a huddle with a Spanish official who had advised during the events which followed the spontaneous explosion and burning of the tanker *Urquiola* off Corunna in Spain on 12 May 1976. He had arrived in Brest with a portfolio of photographs of the affair and talked enthusiastically of burning away 1.5 million gallons an hour. In the *Urquiola*, 35,000 tonnes of oil was said to have been burned, 15,000 tonnes cleaned off the coastline (using 1,200 tonnes of the more toxic first and second generation dispersants) and 50,000 tonnes was said to have been dispersed naturally by winds and tides. Amoco made their own checks, for instance with the US Coastguard, which brought Rinkema to the view that salvage experts still considered burning 'a very controversial subject'. He said 'There are serious doubts if any burning under controlled conditions can be carried out efficiently and successfully' and added that it remained his view that the plan to pump the oil was the best hope. But there was still no survey of the area around the wreck and without it they could not move: the rock that had taken the bottom out of the *Amoco Cadiz* was not even on the existing chart, remarked Rinkema grimly to emphasise his point. His own estimate of the oil spill was even more pessimistic than that of Coulondres the previous day, his best guess being that only 50,000 tons of oil now remained aboard her. In spite of the situation, the idea of burning the ship was quietly dropped once more.

In terms of the ship there was nothing to do for the moment but sit and do nothing.

Yet, six days after the grounding of the *Amoco Cadiz*, it seemed POLMAR resources were beginning to come together. Out at sea, in spite of the weather, which meant that most of their crews were seasick most of the time, more boats were joining in the slow-moving spraying operations; five British tugs had begun work bringing the total fleet to twenty-eight vessels, of which ten or twelve, including tugs, minesweepers, lighters and patrol boats, were on station at any one time. Overhead the little Alouette III helicopters from Lanvéoc-Poulmic continued their activities as spotters, passing their reports down to two French warships, the *Jean Moulin* and the *Détroyat* which were now acting as floating command posts. The Navy reported noticeable breaks in the slick now as they sprayed their chemicals and dumped their chalk, making the spotting of the thickest portions even more difficult, but whether it was due to their actions or the natural processes of the sea seemed impossible to say. Eight hundred and fifty sailors were involved in equipping and crewing the boats or had been loaned to POLMAR-Terre to help with the shore operation. On the coast itself, the clean-up operation was slowly and painstakingly being put together. Several hundred people, including a growing number of local volunteers among the khaki impermeables of the Army, the yellow oilskins of the Navy and the black capes of the gendarmes, were now helping, but their equipment was still pathetically outstripped by the ever growing and moving slick. POLMAR had been launched in the Côtes du Nord during the day as the oil moved east. And it was also being put into action in the next department along the coast, Ille-et-Vilaine, which included Dinard, the holiday resort where Yvon Bourges, the Minister of Defence, was mayor: a major operation was being planned to bulldoze back the sand of the famous holiday beaches of St Malo, and to move oysters from the beds at nearby Cancale back to the inland side of the hydro-electric barrage across the River Rance between St Malo and Dinard. The oil had not reached Ille-et-Vilaine but no one was prepared to take the chance that it might not.

Along the worst hit parts of the coast, the first of 2,000 fresh troops from the Third Defence Region were beginning to move in, by speeding jeep and truck convoys, with yellow headlights blazing and accompanied by motorcycle outriders, though they had not yet been committed to the fight. The front line was still being spearheaded by the farmers with their honeywagons, working round the clock. At Poldhu, a village about two miles west of Roscoff, a team had moved 1,500 tonnes of oil and water melange in one night. And they had been backed up in the past hours by more road tankers and eight mobile pumps at Portsall, Roscoff and Plougasnou. But for every new aid which arrived, a problem seemed to come with it. The sanitation trucks and honeywagons had only been borrowed from their real jobs and they were also badly needed for those duties: public health still had to be protected and the farmers still had to look after their animals. No real replacement rota had been worked out and work teams were now beginning to find their desperately needed equipment disappearing with no immediate chance of finding more. Ensuring continuity of operations was to become a major headache.

The arrival of new resources had done little to ease the concern of the local people. In Brest, six hundred fishermen, mainly from among the thousand who fished from six hundred boats in the Brest and Morlaix areas, met with Jean-Paul Allio, the national general secretary of their union, the CFDT, to draw up detailed demands on the Government to protect their livelihoods. It was an angry meeting which also went over, once again, the worries about the floating barrages and whether the assurances of the government officials on the use of dispersants at sea could be trusted. Their demands assumed that many fishermen might never work again after the oil, and they asked for an early retirement compensation scheme for men over fifty years old. Of all the fishermen in Brittany, those on the northern coast were the most vulnerable to any disaster, natural or man-made. Most fished from small, inshore boats and by traditional methods, relying on the versatility which allowed them to go for different catches, from whiting on handlines to shellfish or seaweed collecting at different times of the year, to make a living. Many had been ashore for a week before the tanker wreck because of the bad weather, and, though some were still valiantly going to sea, well away from the slick and often on round trips up towards the Channel Islands which took sixteen hours of sailing simply to get there and back, they found catches slim. And they were beginning to give up. Albert Le Guen, forty-six, who skippered the eighteen-metre boat *Tante-Fine* said 'Every morning I go down to the shore and I look. And there are days when I do not have the heart to go out. I turn round and go home and look after my little garden. That keeps me busy, a little. . . .'

The fishermen marched from their meeting to the under-prefecture in the city to hand in their claims.

As if as a reminder of its power to threaten the marine environment, the oil reached out for another vulnerable target that day: the little group of islands known as Les Sept Iles, off Trégastel on the Côte de Granit Rose, which are one of the best known seabird sanctuaries in Europe. The islands, with their populations both permanent and transient of puffins, guillemots, gannets, divers, razorbills, shags and gulls had roused a public outcry when they were hit by the *Torrey Canyon* oil eleven years before. According to the French League for the Protection of Birds (LFPO), 25,000 birds had lived there in 1967 and up to eighty per cent of the population of some species had been destroyed. Some of the birds, particularly the auks, were reported to have not yet returned to their pre-1967 numbers.

Already, many hundreds of birds had been found dead or oiled on the coast from the *Amoco Cadiz* oil. Seabirds are always the most visibly affected form of marine life, and an operations centre to collect them had been quickly established in Brest by the Society for the Study and Protection of Nature in Brittany (SEPNB) with a string of subsidary cleaning and collection posts along the coast set up by the LFPO. What worried the ornithologists as they began their studies was that many of the birds were from breeding populations far outside the immediate area, indeed from outside France itself.

Birds had played another poignant part in an incident that day. In the early hours of the morning, the oil had reached the tanks in Roscoff where the Societe Langouste kept lobsters and other crustaceans for sale to the restaurants of France and Europe. The owners had moved the entire stock in seawater-filled wagons only two hours before the oil arrived at 2 am, working though the night. Asked how he had known that the oil was coming, the manager said it was easy. The songbirds had all left the area the day before.

Maundy Thursday, 23 March, exactly one week after the *Amoco Cadiz* had run aground, saw a vital turning point in the way in which the French handled the affair. The sensational treatment of the oil spill by the media both in France and abroad had brought public horror at the column of black oil advancing up the Channel and a great deal of anger and amazement at the seeming inability of the Government to do anything about it. Prime Minister Raymond Barre and his ministers found themselves under pressure where it hurt most, on the question of national pride. The morning's newspapers summed it up: 'The only weapon of the Bretons in the face of catastrophe is improvisation,' stormed a leading article in *Le Figaro*; while others quoted a Brittany mayor, Louis Charles, as saying angrily 'We are no better equipped now than we were in 1967, we are not masters of the situation, we are at the mercy of the currents and winds.' Across the capital, cynical jokes taunted the Government: 'Put Brittany in *your* car', was scrawled on the walls of public buildings; members of the Friends of the Earth invaded the Paris headquarters of Shell and flung an oil-soaked bird on to the boardroom table. And cartoons showed a sign floating on a buoy in the Ushant shipping lanes: 'Danger. Possibility of one tanker wrecking on top of another'. France was shown as a bird, with Brittany the oil-covered and deliberately idiotic-looking head, saying 'We may not know much down here, but we sure know a lot about oil.'

Barre had offered his resignation immediately after the election victory, to give an opportunity for a new Cabinet to be formed, but he had been told very quickly that he was needed in his job. President Giscard d'Estaing had recognised the reality of the voting and was attempting a policy of extending a hand to the Left in the name of 'national unity', and he was not about to change horses in the middle of such a delicate manoeuvre. With that vote of confidence Barre now felt ready to move over the oil. That morning he was due to meet Marc Becam and a group of other Deputies from Brittany who had come to see him about emergency compensation measures, particularly for the fishermen. But Barre had another task for Becam; re-writing the top level arrangements of the Plan he wanted him to take over as full time 'co-ordinator' of POLMAR. Already Barre had ordered his head of civil defence, Christian Gerondeau, down to Brittany to look at organisational problems. Now he wanted some political clout. Becam seemed the perfect choice. Not only was he already involved, but he was a local man. The Bretons might not work easily with a man sent down from Paris but Becam was someone who spoke the language the local leaders understood. And he was a man generally liked by the public; local feeling was running high and if there was going to be trouble, Becam might at least be listened to. Barre had discussed the idea with his cabinet colleagues the previous day and he was ready to put some weapons into Becam's hands. He would have a completely free hand: the Government would take any steps he recommended, either in fighting the oil or in paying compensation.

To that end the Government was already announcing a new provision of nearly £600,000 for an emergency indemnity scheme which would aid the fishermen.

But much more far reaching was the set of announcements Barre was ready to make on preventing anything like the *Amoco Cadiz* happening again. From Easter Monday, no tanker would be allowed to sail closer than seven nautical miles from the French coast. All tankers would be required to radio their presence as they entered territorial waters, together with details of their route, and any difficulties or breakdowns would have to be notified within a radius of fifty miles from the coast. Any tug going to the assistance of a ship in territorial waters would be obliged to inform the French authorities. All tankers going north east up the Ushant shipping lanes would now be required to keep to the very northernmost edge of the lane, certainly within two miles of its seaward side. And no tanker of more than seventy metres would be allowed in the area between Ushant and the mainland. The Navy, which had already been ordered on to a twenty-four-hour sea and air patrol of the lanes, and a twenty-four-hour watch with a rather small radar with a seven to fifteen mile good weather range from Stiff and Créac'h, would now get better equipment and would survey the area until the new navigational control centre was complete on Ushant. The regulations were some of the most stringent ever laid down by a state in peacetime and were exactly the kind of unilateral moves that IMCO had been hoping to prevent.

But it was quite clear where the French felt the fault lay. The text of Barre's scheme was issued to journalists in Brest simultaneously with the announcements in Paris. But the local copies carried one sentence which appeared to be the Government's first official word on who was responsible for the *Amoco Cadiz*. The handout identified Barre with a statement about 'the situation created in Brittany by the wreck of the *Amoco Cadiz* due to the grave negligence of the captain of the vessel, and of the tug it called for assistance.' Barre's office must have belatedly realised what had been said, particularly in view of the fact that Bardari had already been arraigned on criminal charges, and the offending sentence did not appear in verbatim reports of the statement in French newspapers.

Though Becam's appointment had described him as a co-ordinator of operations, the intention was clear: his job was that of an overall commander, a tacit admission that POLMAR as it stood had not worked. He would control the fight both on sea and on land, but there would be a subtle distinction between the two sides of the operation: while Coulondres would be left very much to his own devices with the wreck, POLMAR-Terre was now effectively going to come directly under Becam's command and the organisational control of Gérondeau and Di Chiara. Di Chiara, still chomping on his big cigar, had been down at the advance post at Ploudalmezeau for two days, sizing up the situation and he made the command structure clear when he was asked by a reporter if he was there to dynamise the operation, 'Me?' said Di Chiara, 'I am under the orders of the representative of the Government.'

François Bourgin, the Finisterre prefect, at Quimper, remained nominally head of the POLMAR-Terre co-ordination effort but his role now would become very much a supportive one to Becam. Though it had not been his fault, Barre apparently still held Bourgin responsible for the rough reception he had had at Landeda earlier in the week, and it would not be long before he was replaced as prefect (indeed, it was to become a precarious post; Bourgin's successor would be replaced after local farmers protesting about farm policy hung a live pig from the rotor of Giscard d'Estaing's helicoptor during a Presidential visit).

Becam's new job came none too soon. In Brittany the bad weather conditions of the past few days had worsened with winds first from the north west, then backing to south west and never much less than twenty knots and sometimes up to Force 6. The Navy had been out again to the wreck and found that the bow was tilting up at a sharper angle and that the whole forward section seemed to have shifted on the rocks. The movement had caused a bigger break up of tank three at the rear of the section, and the oil in tank one, near the bow, seemed to be dropping very quickly. All the tanks appeared to be contributing to the flow of oil and the pilot of a Navy helicopter which flew over the hulk reported 'She's vomiting everything she can.' Ashore the story spread quickly through the town centre hotels in Brest, undenied by Amoco and fanned by Jules Legendre, the deputy mayor of Portsall, that all the oil would be out of the wreck by the weekend. It now seemed accepted as inevitable, even if not yet stated, that the pumping operations planned by both the Americans and the French were dead.

Becam had gone on record in answer to a direct question as saying 'So long as there is 10,000 tonnes left in the tanks, we will want to try,' but there was now no longer any strong public defence of the operation. The strong seas and the increasing tides were having a visible effect on the wreck and everyone was now waiting for the big tide of the Sunday evening, known in local seafaring jargon as 'Coefficient 101' to see what that might do. If it lifted the tanker and then dropped her down again, she might part somewhere else along her length. The first thin sheen of her oil slick had been seen approaching within fifty-five miles of Guernsey in the Channel Islands where the Royal Navy and the Department of Trade had set up an advance base for the British operations, and more British boats were being called up. The oil had been spreading towards the centre of the Channel under the influence of the south west winds, and though it had not proved to be the threat to the British mainland that had at first been forecast, the islands, huddled against the Cotentin Peninsula of France, appeared directly in line if the oil should keep moving. On the Cotentin itself, sweeping round from St Malo up towards Cherbourg, oyster growers had already started to move their stock and POLMAR had been set in motion in the Manche department. The Bretons were horrified that the oil might reach the beautiful island of Mont St Michel, east of St Malo, with its historic abbey perched high on a towering rock pinnacle. Somehow it seemed to symbolise their own situation: a beautiful part of France, isolated but attached (a causeway ran between the mount and the mainland) surrounded by an invading force with not enough weapons with which to fight.

'Think what kind of action would be taken if it was a wagon of nitroglycerine near Montparnasse or the Champs-Elysées,' wrote François-Régis Hutin in an Easter editorial in *Ouest France*, 'at sea the risk is just as considerable . . . these tankers and their cargoes are a risk to entire populations . . . the sea is in great danger: will France refuse to heed the cry of alarm?.'

In the streets of Brest, a massive demonstration showed the continuing anger of the people. Six thousand men and women, most of them young and led by the university students, marched on the Marine Prefecture. Older people, watching from the pavement, nodded in sympathy and shouted messages of support. The marchers poured across the tree lined square in front of the eighteenth century fort and in through the portcullis gate of the central courtyard. As Navy staff slammed and locked the main glass doors of the prefecture some of the crowd began trying to tear the massive wooden portcullis from its mountings, while their leaders made speeches through battery-powered loudhailers. A Navy officer who tried to cross the courtyard to his waiting car was jostled and pushed and a balloon containing oil was thrown at him, splattering his uniform. The marchers then staged a sit down in the courtyard and remained until the Navy agreed to accept a protest note through the front doors. A cheer went up from the crowd as the doors opened and some of their leaders were allowed inside and the assault on the portcullis stopped. When the leaders re-emerged the crowd started to melt away through the gates in the weak early evening sunshine and back into the town.

The police had wisely stayed away completely so as to avoid a confrontation. But as the marchers left a Navy fire tender was drawn up outside the gate, with its water cannon at the ready.

By the next morning, Good Friday, 24 March, it was clear that the sea was working on the wreck with a terrible strength. When the tanker had broken its back in the first few hours after grounding, the two parts of the ship had been held together by distorted metal on the port side. But when the Navy carried out its usual first light inspection, they found that the two halves had now been completely torn apart. The rear portion, with its towering 'chateau', the massive 8-cylinder, 30,000 horsepower diesel engine, and the huge funnel with its defiant Amoco symbol, had been picked up and swung bodily round through ninety degrees, from pointing south west to south east. Much more oil seemed to be escaping as a result. If one night's seas could do that there now did not seem to be much hope that the rest of the wreck would not be punished. How much oil remained inside her could only be guessed at, no one could board her with twenty-five foot waves crashing across her deck.

The overflights now showed that the oil slick stretched more than eighty miles along the coast from Portsall to the Ile de Brehat, off Paimpol, and at least eighteen miles out to sea. The thickest deposits were still plastered into west-facing areas, particularly around the Ile de Batz, off Roscoff, and in the Bay of Lannion from St Michel-en-Grève to Trégastel.

Heavy slicks were scattered across the Bay of Morlaix, where the teams who had fought all week with barrages to keep the oil out of the estuary and the extensive oyster beds now saw that they were facing a losing battle. Most of the thick oil had been kept outside the booms, but the aerial infra-red photographs showed a thin layer reaching inland up the estuary, over the beds and almost up to the big town of Morlaix itself.

With Becam's appointment and the day of the big tide fast approaching, the *Amoco Cadiz* operation was beginning to move towards something of a climax. Whatever the state of the wreck, POLMAR strategy was still for a once-and-for-all clean-up following 'Coefficient 101'. A date had been set for the start of the work, Tuesday 28 March, the first day after the holiday weekend, and the military had even coined an unofficial title for it: 'Operation Teaspoon'. Becam's first move was to get his lines of command established. For this he, Gérondeau and Di Chiara moved into the advance post at Ploudalmézeau. From here, not only would they be at the battle front, but the civil defence men could liaise directly with local co-ordinators and with CODISC and CICOPH in Paris, to get what they wanted.

The setting up of such a post had not been foreseen in the original POLMAR plan, much less its control by central government, but its role had become essential. Saint-Prix, the under prefect of Morlaix, had had the post in operation since the Sunday and found himself immediately inundated with demands and propositions for fighting the oil. A plethora of equipment, most of it useless had started to build up as the first, unco-ordinated arrangements were made, and Saint-Prix had taken it as his main task to bring some kind of order. It had been no easy job, for, as Prefet Bourgin was to reflect later, there were up to eighteen different administrations involved, not one of which seemed to want to move without checking back with its own headquarters and each of which appeared to have its own ideas about what should be done. Communications had been a major difficulty; sixteen extra telephone lines had been laid on by the Post Office, but they were still not enough. Officials, working twelve to a room, sometimes had to go out to public call boxes if their calls were urgent. At some times of the day the only way military officers there could get in touch with their units was through the radio of a helicopter standing next to the control post. Some of the problems had been solved by bringing up a second and bigger command vehicle from Rennes to sit next to the first and offer radio communications with all sectors; but to the casual visitor the post still seemed to be in semi-organised bedlam, an uninterrupted ballet of gold braid, silver stars, green and black berets and blue kepis mingling with suits and shirtsleeves as the different organisations went about their work. Forty people worked at the post, divided into separate cells for information, communications, civil and military means. This organisation would continue with one change: Ploudalmézeau would in future continue to be the advance post only for Finisterre. Not only was that where most of the oil still was, but the launching of POLMAR in the Côtes du Nord meant they could take over responsibility for their own beaches.

Indeed, based on a headquarters in the prefecture at St Brieuc, with an advance post at Lannion, the Côtes du Nord organisation escaped some of the criticism of the hastily constructed Finisterre system, by being built round the local administrative and political machine. With an initial response based on 150 local firemen split into three flying columns, and then the farmers with their honeywagons, the front line sectors were to be overseen by sub-prefects and local mayors were given a specific liaison role with the local people, kept in touch with what was happening by meetings and daily news bulletins.

The second part of Becam's task would be to ensure that POLMAR secured the means for Operation Teaspoon and put them into action. Already more than 300 machines of different kinds were on the shoreline, including the honeywagons, sanitation wagons, trucks, pumps, tankers, public works and military vehicles. But progress was still debatable. Official figures suggested the amount of material being shifted had gone up from 250 tonnes a day on the first weekend to 2,000 tonnes now. But other figures from Brest, where the oil and water was being pumped into barges and then across to a treatment plant for separation, did not back that up: the figures there estimated only 1,500 tonnes of the melange had been brought to them in four days, of which only forty per cent was truly oil. More vehicles would be needed particularly to transport material, but on a coast which varied from saltmarsh to granite reef, mechanical means could only provide a part of the answer; access was impossible in many areas, and many of the vehicles could only operate on hard surfaces; thousands of square feet of metal grills were being brought in to stop wheels ploughing into mud and sand.

While the vehicles would provide support and work in bay and port areas, what would be needed would be sheer manpower, down to the humble bucket brigade. There were now 450 soldiers in Finisterre and 500 in the Côtes du Nord, with five companies in reserve; then there were the seventy UISC7 civil defence men, seventy-five marine commandos who were in charge of boom operations and other Navy personnel; together with civilian workers, including the farmers, who probably added up to one thousand more. But the response was still out of scale with the problem when the oil was spread along eighty miles of shore. That was measured as the crow flies, but the Brittany coast was one of the most crenulated, inlet-ridden and embayed in Europe – very similar to the shores of northern Maine and parts of southern Alaska – which meant that the actual mileage affected by the oil might be double that or even more. Even if the remotest rocks were left for the sea to scour clean, it was clear that thousands of pairs of hands would be needed to help. The decision not to use chemicals inshore and the weather, which forbade more extensive mechanical means, left no other choice.

Marc Becam went on radio and summed up the situation: 'We are not without powers, but the disaster is of such a size and we have had to come to terms with the weather.' More troops and manpower from the civilian administrations could be called in. But POLMAR officials had been handed another, somewhat double-edged sword. The sensational publicity that had so embarrassed the Government had also had the effect of creating a spontaneous wave of sympathy for Brittany both in France and cross most of Europe.

Thousands upon thousands of people seemed to want to volunteer to join the beach gangs and clear the oil. Fanned by a strident campaign on the commercial radio station Europe 1, which broadcast over most of France, people in dozens of towns were now arriving at their local town halls with heavy rubber gloves, thigh boots, oilskins, dustbins, buckets, spades, rakes and rags for the clean-up. Within six hours, according to organisers of the appeal broadcast under the slogan 'There must be a way to help Brittany,' three times the 6,000 pairs of gloves and 2,000 dustbins called for had been collected, along with 2,300 out of 3,000 oilskins and four tons of rags. Stores and private firms had donated a lot of the equipment. Becam spoke on radio pledging 'national solidarity' in the face of the oil and talked grandly about a European solution to pollution disasters in which each country would support the other, each specialising in a different means of intervention.

But at the same time the idea of an invasion of volunteers posed a major and unforeseen problem. The local people who had volunteered in the first few days after the wreck had been welcome: they were motivated and had a useful knowledge of the coastline; they were also able to return to their own homes each night. But there were absolutely no arrangements for feeding, housing and equipping thousands of people who might pour into the area from all over the nation, much less any scheme for controlling them and putting them to work.

Most of those who wanted to come appeared to be young, not only because the young might be expected to be more concerned about the environment, but because the schools and colleges had been on holiday since 1 March; and there were no arrangements for insuring their safety. Nor could their motivation always be relied on. Already many of those who had inundated officials with telephone calls, telegrams and letters had given the impression they simply wanted a holiday on the state with a few hours of work thrown in. And the officials spoke in derisory fashion of the 'hands in pockets gangs'. The last thing they wanted was a massive influx of disorganised, even if well-meaning individuals, unequipped and uncontrolled, tramping the beaches and generally getting in the way. They had already seen the jams and confusion that a few thousand sightseers had caused in the Portsall area on the previous Sunday. So with the holiday weekend beginning POLMAR found itself applauding the idea of national solidarity with one hand and with the other moving in detachments of gendarmes to police the coast and to keep sightseers and would-be volunteers at bay. The action brought an immediate heated row with local shopkeepers, hoteliers and restaurant owners who, fearing their season would be wiped out by the oil in any case, were glad to see anyone coming to spend money in Brittany. The hoteliers in particular protested that so far they still had firm bookings and they did not want their customers driven away. 'Why should we lose this important weekend trade?' protested Charles Pavot, mayor of Porspoder.

Within hours the controls had had to be relaxed so that those with holiday homes or firm hotel bookings would be allowed through the police cordons. But Pavot was still left shaking his head, 'There is no reason for it', he said. Becam, who had welcomed the appeal for equipment, ('another dustbin given is another dustbin of oil taken away'), did not like the idea of refusing the enthusiasm of the volunteers but saw the problems. And while officials at national and local level tried to work out a plan for handling them, he had to make a new radio appeal for people not to come to Brittany, at least for the clean-up, until Operation Teaspoon got under way after the holiday weekend, and then for them to come only when asked. Becam also appeared anxious that outsiders should not get the impression that France's major reply to a national emergency was to mobilise its school children, and he made it clear that the largest part of the work in the operation would still be handled by the authorities. 'The State will assume its responsibilities,' he said firmly.

The oil was also going to have to be fought at sea and Becam busied himself during the day with the visit from the British shipping Minister,

Clinton Davis, who could provide more spraying vessels. The British open waters defence against oil was based almost entirely on the use of the new generation of dispersants and a great deal of thought had gone into the sprays and agitators attached to their boats, which had been developed at a division of the Government's Warren Springs laboratories, near Lowestoft.

French vessels had had to compromise and British marine biologists and oil spill experts giving advice on shore had been horrified by local newspaper pictures showing dispersants being spread from a minesweeper with fire hoses: the very essence of successful dispersant treatment was the proportion of chemical to seawater and ensuring even distribution. Though the British had now brought in their own command vessel, the frigate *HMS Zulu*, they were also keeping in liaison with COM at Brest to help the French when reinforcements were needed, and over the weekend thirty French and British vessels would be attacking the oil at one stage, as a preliminary to the clean up operation. But the meeting with Davis was also politically opportune for rumours had been spreading that the British were becoming increasingly exasperated with French tactics, particularly with not having moved earlier to begin clearing up the oil ashore. Whatever the truth Becam and Davis made a point of looking relaxed and smiling together and Davis reinforced the point by having his Department of Trade send a telegram to the French denying any friction, saying that the two countries had worked cordially together, and asking for closer future relationships. But at his press conference Davis avoided saying whether or not he approved of the unilateral moves the French had made over Channel navigation for tankers: the move would, of course, have had the effect of moving such traffic nearer to the English coast. Davis reserved himself to saying that the subject would be raised at Common Market and IMCO meetings later in the year.

The show of cordiality did not impress everyone. That evening 3,000 demonstrators from unions and conservation groups took to the streets of Brest in a protest more ugly than the day before. The protestors marched on the under prefecture, which was now guarded by police. Fifty oil-soaked birds were thrown into the prefecture courtyard as demonstrators tried to scale and pull down the surrounding fences and the police fired tear gas into the yelling crowd, which broke up and then assembled in the main Rue de Siam to make for the town hall. As the crowd made its way a rumour began that some of the marchers had been arrested and taken to the police station and a section of the crowd broke off and ran to the Avenue Clémenceau intending to rescue their colleagues. They were headed off by Louis le

Roux, the deputy mayor of Brest, who had taken part in the first and orderly part of the march, who went inside on their behalf and established that no one had been arrested. For some time the crowd occupied the square in front of the town hall, smashing the windscreen of a police car and daubing it with red paint, and stoning a police van, before dispersing near to midnight.

On Easter Saturday 25 March, Smit International managed to get their own expert aboard the *Amoco Cadiz* again, and he came back with grim news. Just as had been feared, the sea had been working on her dreadfully and she was now ready to break apart again, this time between tank two and tank three, and was spilling a great deal more oil. Tank one was almost full of water and no more than 35,000 to 38,000 tonnes of oil could remain in the wreck at the most optimistic. There was probably as little as 25,000 tonnes.

Rinkema and the Amoco team went into a crucial meeting with the French to decide what to do, broke overnight to talk among themselves, and then on the morning of Easter Sunday, the day of 'Co-efficient 101', resumed to make a fateful decision; that the pumping operation would finally be abandoned.

There was now only one course of action left and that was to evacuate the wreck of oil as quickly and as simply as possible so that it did not become a long term pollution hazard and let all the remaining cargo be handled by the clean up teams. After consulting Shell and the insurance syndicate, who technically still owned the hulk and her remaining oil, Rinkema gave permission for the French to act. Just after noon, Navy divers went out to the *Amoco Cadiz*, her decks awash, and began to open fifteen deck access hatches on the oil tanks, pipes, vents and any other openings which would give the sea access to the oil. When the big tide came in the darkness of the evening – and there would be a difference of as much as thirty-five feet between high and low tide with perhaps half as much as that in swell – the force of the water would ram oil and water out of the hatches and openings like gigantic geysers.

Even during the afternoon, helicopter flights around the wreck could see columns of water twenty feet high showering upwards from the deck and being torn away by the wind. Marc Becam said he had approved the Navy plan to 'Free all the oil as brutally and as quickly as possible'. The Navy would finish the job, probably the next day if the weather was right, by sending in the divers again to place demolition charges around the hull and blast holes in any tanks still suspected of holding oil. Becam said the pumping operation had failed not only because of the weather but because of the small area of the tanker now above water and the danger from gas to the salvage crews.

It was a bad and dispiriting day for the Amoco party. The French had made it abundantly clear to them during their stay in France that they and their company were to be held directly responsible for the pollution. And their captain had been charged with negligently polluting the seas. Now Rinkema and his experts had put forward what they saw as the only practical solution to minimise the damage from the situation only to have it fail. And at the same time they had seen the French stubbornly holding on to their own ideas on what to do with the oil which had spilled and loathe to take advice from anyone who had suggested different tactics. Rinkema had remained totally diplomatic throughout his public appearances and he remained so now, referring only to 'minor friction' with the French authorities who, he said, would naturally have wanted to do things their own way. But it was clear that he and his party had often been exasperated with the workings of the French system. Nor did the conditions under which they were working at POLMAR-Mer headquarters lead to an easy understanding: the subterranean conference room sometimes held forty-five people with something to say.

But while Rinkema remained silent over the French strategy, there were other voices making themselves heard forcibly. During the first week of the spill a growing number of companies, particularly from America, specialising in treating oil spills, had flown in men and equipment to offer to the French. The offers were not altruistic, lucrative contracts were envisaged, but the companies saw themselves as experts in the state of the art and had become angry when their advice, and their products, were largely ignored. One firm managed to sell a batch of thirty-five small compressed air-powered skimmers which could be used by the shoreline teams, and another with a machine using a continuous belt to mop oil off water had been given a contract to work in harbours and inlets, and both proved effective. But most firms complained they faced a labrynth of bureaucracy in which they were never able to reach anyone who could take the decision to listen. And they were unanimous in their condemnation of the French. 'When a city block is on fire, you don't start choosing your pumps,' exploded one, angrily. All seemed agreed that the French should have begun a large scale action to clean the beaches on the first day, using anything and everything that came to hand, whatever the state of the tides. But the French for their part remained dismissive of this kind of expert advice. They remembered the Dutch company who had offered to handle the entire spill for £2.8 million, and the Norwegian company which was

Soldiers collect oil-soaked weed from a beach just north of Portsall.

A group of workers struggled with the heart-breaking task of cleaning a rocky section of the coast.

even dearer. 'I am always here for anyone who wants to see me', smiled Christian Gérondeau disingenuously from his Ploudalmezeau base. And Becam himself retorted that while he welcomed any help, he 'would not be prepared to pay out £1,200 a day for an American boat which was manifestly unable to work in waves over a metre high.'

Indeed, the uselessness of a great deal of advanced technology in the face of the sea and weather conditions became a marked factor in limiting even official international aid in the initial days of the fight against the oil. Several countries had sent experts of various kinds, from coastguard officers to scientists who had been able to offer advice (but many of them had been there as much to learn, better to protect their own countries in future, as to help and some frankly described themselves as simply observers); but in equipment and manpower only the British appeared to have mobilised a practical major effort. Most of the hardware from other countries, with the exception of barrages, vehicles and some absorbents for spreading on the oil, had been restricted to individual mechanical oil recovery vessels from major ports, which at first found themselves in difficulties, either because of rough seas or because they were too big to be used in shallow coastal inlets. Only a handful of countries, such as Belgium, which had sent eighteen civil defence workers, and Denmark, which had fielded a fire brigade unit, had supplied any kind of official manpower. The situation was a long way from Becam's notion of a pan-European anti-pollution team.

The French had become very sensitive to criticism of their operation and Becam rounded on one interviewer with the angry retort that 'The Government's effort is neither derisory nor second rate.' And they were particularly ruffled by the continuing strong reaction from some of their own people. For in spite of Prime Minister Barre's hopes that Becam's appointment might go some way to assuaging the local anger, the Easter weekend was destined to become more violent than ever.

First, the most extremist of the separatist organisations, the Breton Revolutionary Army (ARB) began to move.

The *Amoco Cadiz* catastrophe was seen by the ARB and its sister organisation, the Breton Liberation Front, as a typical example of the abuse by capitalism of its powers and another example of the neglect by central Government of Brittany as a province.

And in the early hours of Easter Sunday morning, Shell France's delivery control centre for Brittany, a typical symbolic target for the group, was blasted to pieces by a bomb. The building was mercifully empty at the time but the explosion was just one of a series of similar incidents which would be carried out in the coming weeks expressing their disgust at the *marée noire*, and the reaction of the Paris Government to it, including bomb hoaxes, petrol bomb attacks and more dynamiting. Late in May it was considered still too dangerous for President Giscard d'Estaing to visit the area on a public relations trip to see the results of Operation Teaspoon, and, as if to prove the point, on the day the cancellation of his trip was announced, the ARB struck again, using plastic explosives to blow up communications equipment at a gendarmerie headquarters in St Brieuc. It was the eighth attack since the beginning of the month and the twentieth of 1978.

The attacks reached a climax in June when a devastating bombing, hailed as the most spectacular action by a separatist group since the Second World War, blasted ten rooms in the Palace of Versailles causing more than £500,000 worth of damage. The target, the Napoleonic Galleries in the south wing, dedicated since 1837, in Louis Philippe's time, to 'all the glories of France', particularly the Napoleonic ones, housed all the pictures and sculptures bought by emperors to glorify military history. Only a month before part of the wing had been reopened by the President after years of painstaking restoration. It had a poignant significance since Napoleon represented all the relentless centralism that the separatists professed to hate. But it was a step too far for the ARB-FLB for it brought a backlash from the national authorities in the form of a massive police operation headed by Roger le Taillanter, head of Rennes CID, which tracked down and arrested twelve of their number. Two of them, an unemployed printer and a delivery man, were charged with the Versailles bomb attack and later sentenced to fifteen years each. De Taillanter claimed he had arrested the leaders and all the active members of the movement except for a small cell in Finisterre and, he said, the end was near for that, too.

The second irritant for the POLMAR leaders during the Easter weekend was a pirate radio station which kept appearing on the air with anti-Government propaganda, defying all efforts by the authorities to jam it: the broadcasts ridiculed the POLMAR efforts.

And in the streets the demonstrations began again. During Sunday hundreds went on the march in Brest, Lorient, Lannion and Paimpol. But the most violent events came after a massive march through Brest the next day. It was by far the biggest demonstration organised against the *marée noire* and more than 15,000 banner waving workers and students marched ten to fifteen abreast in the rain down the main Rue de Siam towards the marine prefecture, and back round the town. Loud hailers led chanting from teachers, fishermen, factory hands, lorry drivers, municipal workers, and workers from the Brest naval yards, who had all united under the slogan 'The polluters must pay'. The main part of the demonstration passed off peaceably, with no police presence, but as the crowds started to melt away, the more militant factions remained, milling in the streets and squares, where they suddenly discovered that strong detachments of the paramilitary CRS had been quietly standing by in full riot gear of combat suits, steel helmets, clubs and perspex shields. Running street battles then developed through the town, with the police using their shields and tear gas grenades to protect themselves against fusilades of bricks and paving cobbles, and the demonstrators retreating in the face of repeated baton charges. The disciplined organisation of the police soon broke up the groups of demonstrators, but not before people on both sides had been injured, and two English journalists were almost killed when a CRS man threw a brick through the windscreen of their car. It would be a week of protest in Brittany and the CRS would be out again in force as a frustrated government tried to keep order.

The future of the *Amoco Cadiz* now rested firmly in the hands of Admiral Coulondres and the French Navy. But even now that she had been finally abandoned it seemed as if she was ready to combine with the elements to defy the best efforts to finish her off. The weather, which had been the bane of the whole operation, had been bad over the weekend. On Easter Monday, and despite a heavy running swell, divers had been sent out to her to find out what had happened as a result of opening the hatches and the full flood of Coefficient 101, the high spring tide. The wreck was surrounded by fresh black oil and as the second break in her hull tore further apart, the tanker's bow had reared a defiant forty-five degrees into the air. With her huge bulbous nose, normally underwater, jutting skywards she reminded the divers forcibly of a huge dying whale. They had orders to place demolition charges round her hull and to blast open a pathway for any remaining oil that might still be clinging in her tanks. But after only ten minutes in the surging water they were forced to abandon the attempt at the risk of their lives. Next morning, Tuesday 28 March, the day chosen for the start of Operation Teaspoon, the divers went out again. Now the tilt of the tanker's deck was too steep for them to work on; the middle of the forward section of the wreck had collapsed several feet and the rear had been bodily lifted almost twenty feet by the powerful overnight tide. Though calmer, the turbulence was still too violent for the men to work underwater.

On the morning of Wednesday 29 March, the Navy decided to change its tactics. Overflights of the wreck showed that during the night the forward section had been completely torn apart so that the ship now lay broken into three separate pieces. All three were moving in the sea and there was no way in which the men could be allowed to risk their lives under her.

Coulondres issued his orders and exactly on 15.00 hours, as the sea around the *Amoco Cadiz* was circled by Navy patrol boats and a spotter plane hung overhead, three submarine-killer Super Frelon helicopters from the base at Lanvéoc-Poulmic came up over the grey horizon. In dramatic contrast to their search and rescue role on the night of the grounding the machines now began to carry out the kind of job for which they had been designed: in line astern they thudded low over the wreck and started to depth charge her. One after the other in a practised routine that lasted forty-five minutes, the helicopters dropped a total of twelve mark 56 anti-submarine grenades, each containing 350lbs of high explosives, in a pre-determined pattern across the hulk. Set to go off twenty-four feet underwater, ten of the twelve charges exploded with huge fountains of spray and a power that shook the ground under the feet of those watching from the shore more than a mile away.

Early the next morning, in calmer seas, twelve frogmen went out in two Zodiac dinghies to find out what had happened. Though there was fresh oil in the water all round the wreck, no more could be seen coming from the forward section. The deck had been split wide open though it was impossible to tell if this was caused by the grenades or the action of the sea.

The frogmen, with tremendous courage, dived nearly eighty feet into the inky waters to try to ascertain the damage to the wreck: the first time they had been able to go deep under her since two days after the grounding. Their orders were to find out if any oil remained in the forward section, but it seemed now that it was empty. What worried the divers was the rear section, which contained the two tanks full of bunker fuel and the slop tanks. Was there oil still in there? Their inspection was inconclusive and when they made their report, Coulondres was unwilling to take any chances. In a repeat of the previous day's operation, another flight of Super Frelons was sent up over the *Amoco Cadiz* to plummet their grenades on the 'chateau' of the stern section. The suspicions of the frogmen had proved correct and the helicopter pilots now saw 'an enormous ball of oil' surge to the surface from one apparently unbroken tank and start to move away towards the coast. Exactly two weeks after she had gone aground, the *Amoco Cadiz* had put her entire cargo of 223,000 tons of crude oil into the ocean and lay torn to pieces on the Atlantic reef. In another eleven months she would disappear forever beneath the waves in a storm very much like the one that had delivered her to the rocks.

9 / Operation Teaspoon

THE CLIMAX OF THE *Amoco Cadiz* affair came with Operation Teaspoon. After nearly two weeks of waiting and preparing the French were as ready as they would ever be to attack the spill on a large scale. But even now the oil was not ready to make their job any easier. In the two weeks after the grounding, as we have seen, all 223,000 tons of the tanker's cargo had gone into the ocean. It had been swept eastwards by winds averaging twenty to thirty knots and tides and currents which pushed and pulled it into a broad band, generally eight to fifteen miles wide, along eighty miles of coastline until it reached the Ile de Brehat on the brink of St Brieuc Bay. And there it had come to a halt. The elements had been fickle. Some of the shore was covered with hundreds, sometimes thousands of tons of oil, while other parts were hardly tainted at all. Some areas, small coves, inlets, sandy spits between islets and land, flats and salt marshes, seemed to act as natural reservoirs and sinks for the oil, attracting tremendous amounts; while elsewhere the sea deposited the oil only to return to sweep it away again. At Brehat, great iridescent tongues had reached out more than forty miles to the north east towards the Channel Islands: thin oil making a final defiant gesture which still kept the French and British spraying boats on watch in case thicker deposits were to follow.

But as March turned into April, and the big clean up operation got under way, a new phenomenon appeared: the winds, which had blown steadfastly from a westerly direction since the grounding, turned about and began to blow just as consistently from the east. And the oil began to move again, back from where it had come. For at least a week, but then again later, to the dismay of those ashore, the oil pushed resolutely westward, past the wreck that had spilled it, and then south down the coast of Finisterre as far as the Bay of Douarnenez, over two hundred miles from Brehat. Thinner this time, the Navy reported it floating out at sea in patches thirty to forty yards across. Now more tenacious and widespread, it flowed back on to rocks and beaches already polluted and sections it had not previously touched at all. In total, more than 245 miles (far more if the indentations of the coastline were taken into consideration) were affected, to some degree or another.

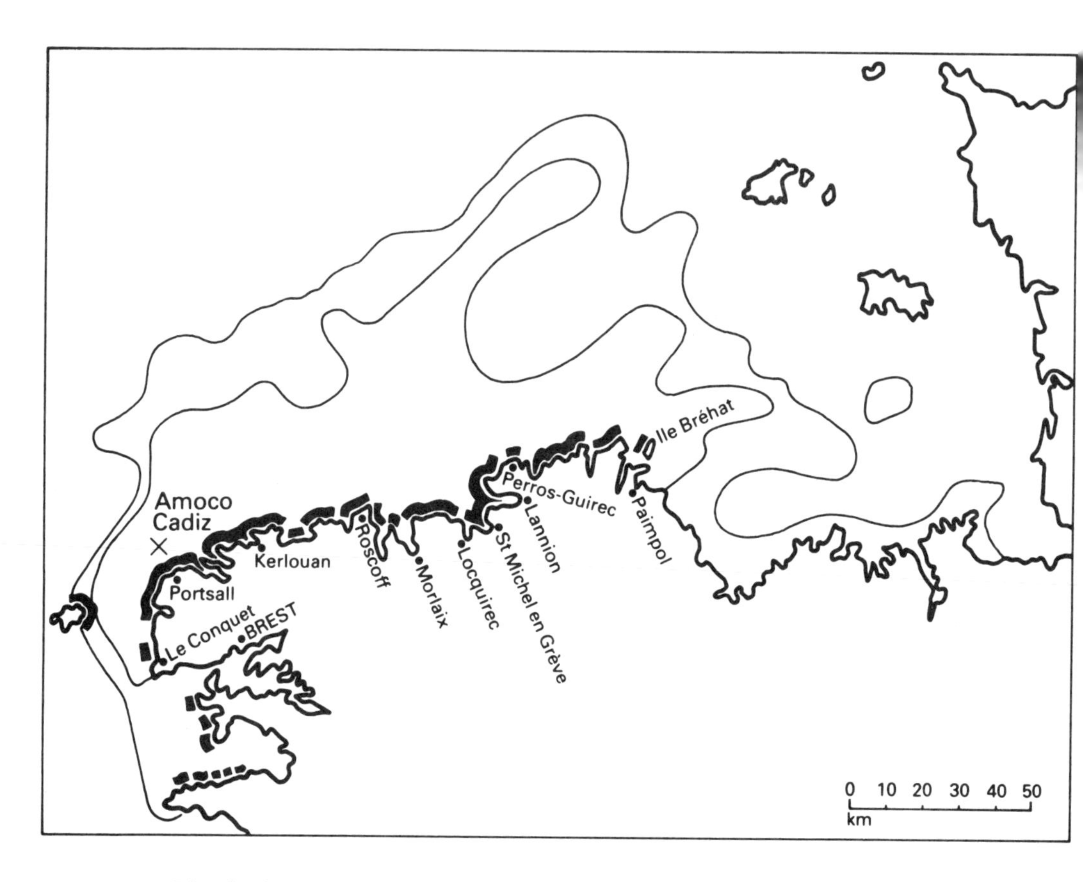

The final extent of the coastal pollution and the slicks out at sea. The inner line indicates the area covered by heavy floating oil and the outer the spread of the lighter sheen. The heavy lines mark the extent of the stranded oil.

The theoretical scale of what the French were attempting to do was gigantic. A group of American scientists from Texas A & M University, working for the National Oceanic and Atmospheric Administration/US Coastguard Spill Oil Research Team, later tried to put a figure on how much material might have been waiting on the coastline to be shifted and how much transport it might have taken to do it. Their calculation assumed that as much as thirty per cent of the total cargo, or 74,000 tons, representing the lighter fractions, had already disappeared, either by evaporation into the atmosphere or by dissolution in the water column; and that all that remained, 149,000 tons, had turned into mousse with the potential of coming ashore. Adding the weight of the combined water to the oil, thought the Americans, might have meant 488,000 tons of mousse waiting to be cleaned off the coast, or over twice the original weight of the oil cargo. And that did not include the potential extra volume and weight of any sand, weed or debris which got mixed with the oil and which might represent up to twenty times the oil content in the resulting mixture. Looked at another way, there might have been 18,940,000 cubic feet of material, 'capable of coating 359 miles of coastline with a layer of mousse one inch thick and 120 feet wide.' As for transport, taking vehicles they knew had been called in for Operation Teaspoon and beginning with the biggest, they calculated that everything could be shifted in 17,702 loads using 8,000-gallon railway tankers. But railway tankers plainly could not be parked behind each beach, and further calculations examining the more realistic forms of transport illustrated more clearly the theoretical enormity of the operation: it would take 35,405 loads using 4,000-gallon road tankers; 75,170 loads using 252-cubic foot trucks; 78,678 loads using the 1,800-gallon tankers like those in action at Portsall; and a massive 118,830 loads using the 750-gallon honeywagons, which were often the only vehicles that could get on to the beaches.

The French remained optimistic. 'I'll bet you a bottle of champagne', said a senior official, 'that you can come back in June and sit on a perfect white beach to stare at what's left of the *Amoco Cadiz*'. And Michel d'Ornano, the Environment Minister, was happy to boast: 'Before summer, all this will be just a bad memory.' And certainly the resources began to pour into Brittany. Though manpower figures remain confused, by 3 April there were said to be 5,800 military personnel on the scene: 2,000 Navy, 2,800 Army and 1,000 gendarmerie; together with a flight of Super Frelons and a squadron of Alouette helicopters, a flight of Alizé light aircraft, 75[illegible] commandos and 250 marines from Lorient. The Navy had 21 boats at sea with 735 men. And in addition there were firemen, civil defence workers, ministry teams and volunteers said to top the 3,000 mark; and 1,008 vehicles including 300 military, 108 fire engines and pumps, 125 tankers, 80 honeywagons, 170 public works vehicles, 72 sanitation wagons, 103 skip lorries, 50 large trucks and 60 tippers.

When the oil first came ashore from the *Amoco Cadiz* it had been light and liquid, easily dispersed in the churning water and even more readily turned from its pure jet black form into a reddish brown water-oil emulsion or mousse, and in some conditions even into a chocolate brown froth the scientists had never seen before and which they dubbed 'mousse-shake'. It had spread quickly and easily across the water, coming in and out on the tide at first, and then leaving large amounts when the sea ebbed, pouring into crevices, pools and natural hollows up to a foot deep, or sitting in great sheets on the sand with the highest concentrations in the high tide zone. Each successive tide would pick up the stranded oil and carry it further up the beach. A typical beach would have a fifty yard wide band of mousse left stranded at the top of the intertidal zone which would either stay in place, or ooze back down the beach face if it was steep enough. Down the beach, the oil would be floated back to the sea's edge by rivulets of ground water and then come back in with the next tide. It would bridge the fine sands, rather than sink in and in quiet corners away from tidal action, large reservoirs of oil would gather, unmoved by the sea.

But as Operation Teaspoon began work, the oil from the first spills had begun to weather and age: the lighter fractions evaporated and the remaining oil became more viscous, picking up sand, seaweed and debris, and resisting the floating and carrying action of the water. More of it began to stay behind on the beach or marooned on tidal flats and salt marsh. Even while some of the more exposed sections of the coast were being scoured of oil by the ocean itself, the falling tides after the spring equinox were leaving more and more material at high tide mark. In the months between March and June a great deal of sand and sediment was normally suspended in the water, (this particular year there was more than previously) and the action of the water was to scour the beaches and then to lay down new layers of sand with each incoming tide.

In this way, though there was still fluid mousse in the sea, a great deal more was becoming something of a sticky semi-solid melange, or was buried in layers of beach sand like a mille-feuille cake, sometimes as deep as twenty-five to thirty inches. A tendency of the fresh mousse to stick only to dry surfaces was being replaced with a glue-like tenacity everywhere, particularly in rocks and weed. And what had started out as a floating, pumpable liquid was undergoing metamorphosis into a stable, heavy, non-pumpable material that could only be shifted by slow and laborious physical removal.

It is not clear whether anyone truly knew when they began what it would take to handle the problem or whether the massive teams of men and machinery being brought in could be co-ordinated and put to work without organisational disaster. But in scenes reminiscent of a full-scale wartime invasion, convoys of Army trucks began to haul the first of thousands more troops up to the beaches; police guided lumbering tankers, jeeps and officers' cars with pennants fluttering; civilian vehicles including pumps, trucks, tankers, fire engines, sanitation wagons, tippers, and tractors with the ubiquitous honeywagons in tow roared to and fro along the narrow coastal roads; trains of railway tankers were backed into local sidings; overhead, helicopters and light planes weaved and circled; and on the shore itself, the first of thousands of young volunteers mingled with the soldiers, gendarmes, marines, sailors, civil defence workers, firemen and government teams in a kaleidoscope of different coloured oilskins as the 'Chinese coolie' brigades, as the local newspapers called them, began to attack the oil once and for all.

Up to 29 May, said the official inquiry reports later, the average was 72 teams made up of 4,600 soldiers and other military personnel, 150 firemen and 750 volunteers, though other sources show a manpower peak in mid-April of nearly 6,500 made up of 4,500 military and more than 2,000 volunteers. According to the Texas A & M University team the total manpower for the operation might have been as high as 8,187 with two men in Finisterre for every one working in the Côtes du Nord.

The clean up began in an atmosphere of official determination from the French, as expressed in a number of separate incidents over a few days. Four tankers which had been hoping surreptitiously to wash out their tanks in the Channel had been caught and dealt with by the newly vigilant sea and air patrols. In three days, they found the *Sunny Queen* (Norwegian) north of Ushant; *Le Costas* (Greek), west of Guernsey; *Gogo* (Liberian) north of Les Sept Iles, and *L'Akto* (Russian) in the Pas de Calais. To make the point the French ordered *L'Akto* into Dieppe and fined her master £3,571, which brought a stinging rebuke from Tass, the Russian state news agency that the French were seeking scapegoats for their own culpability with the *Amoco Cadiz*.

At the French port of Donges, the 113,000-ton British tanker *Nordic Commander*, which had put in with boiler trouble, was immediately boarded by a Marine Affairs Department officer and ordered not to sail until full repairs were complete. And the Navy arrested the French ship *Le Gers*, on its way from Le Havre to Bordeaux, after it had sailed through the Chenal Four between Ushant and the mainland, despite orders not to do so. And as if to seal the point, on 4 April, Captain Harmut Weinert of the *Pacific* was taken before Judge Gouyette once more and formally charged in the same terms as Captain Bardari with polluting the seas. He was released on bail.

The clean up began with enormously high morale among the workers. Columns of laughing young soldiers clowned in their ill-fitting oilskins as they were marched down to the beaches: 'See,' shouted one, pirouetting his cape, 'A *la Chinois*! ;' 'If I hadn't been ordered here as a soldier, I would have come anyway,' said another crew-cutted private from Rennes, to nods of approval from his comrades. Officers and civilian beachmasters, mostly from the Ministry of Equipment, deployed the men and gloves, buckets, spades and rakes were issued from dumps established along the coast. One equipment base at a fire station at Kerlou, near Brest, held more than five hundred tonnes of these simple implements, some of them the result of the radio appeals.

Work gangs were issued with instructions from the anti-poisons centre at Rennes on handling the oil. They were to keep on their rubber gloves at all times and to wear goggles if possible; they could expect frequent complaints of nausea, headaches, digestive upsets and diarrhoea. Those suffering from bronchitis or other chest complaints should not take part. And they should avoid confined spaces: the heavy fumes could produce a nervous excitability very like being drunk on alcohol which could be dangerous when working with machinery. Anyone who fell into the oil should wash his eyes immediately with clean water, and if he swallowed oil should drink a milk emulsion and go straight to hospital for observation.

With the oil spread along such a length and variety of coastline including sand, shingle and cobble beaches, rocks, inlets, rivers, creeks, estuaries, islands, harbours, tidal flats, shallows and salt marshes, all of them hit to different degrees, Operation Teaspoon found itself having to respond with a plethora of different cleaning techniques. For some they were able to seek advice, for instance from local marine biologists who knew the coast well; for many others it was simply a case of learning as they went along.

Where the oil was still fluid enough to be pumped, for example at Portsall and at Roscoff, it was attacked by vacuum trucks and honeywagons, the trucks standing on jetties and ramps or on prefabricated roads of metal grids, and the wagons following the sea as the tide went in and out. Some spots had some of the small skimmers bought from the Americans, but where they were unavailable, weed was a problem and the mousse was decanted into twenty foot skips hung with wire baskets to strain the weed and debris, and then pumped into tankers. Alongside the vehicles, long columns of men, often standing up to their thighs in the water, manhandled suction hoses or skimmed mousse straight from the water's surface with buckets to be dumped into thirty-gallon drums and then sucked up by the vacuum trucks.

Trenches were dug in the shore, parallel to the sea to retain the mousse as the tide fell. In the *abers*, the continuous belt oil mops were working, and some small mechanical skimmer boats hired privately by the owners of the oyster beds, but these could be used only at high tide. Local oyster-fishermen, unable to match this kind of investment, made wooden rafts to which they attached small outboard motors, which enabled them to reach the very shallow waters at the edges of the *abers*, to clean rocks, coves and creeks unapproachable from land.

Where the wide bands of mousse gathered on beaches, a major operation was mounted to remove it, with large gangs of soldiers, often using long-handled wooden squeegee boards, pushing it into specially dug pits or collection trenches, or natural hollows in the beach where it could be pumped. Weed was piled by hand or mechanical means and allowed to drain and then put into 40 kilo sized plastic bags once used for farm fertiliser. Where the mousse had picked up sand and debris and turned paste-like, work teams using spades, and mechanical front loaders shovelled it directly from the beach and dumped it into trucks.

Large amounts of mousse were caught in rocky areas where seaweed, especially near the highwater mark, seemed to act as a natural absorbent and had to be cut away. The degree of mousse formation was so surprising that Dr Jean Vasserot of the Marine Biological Station at Roscoff wondered if the dense covering of algae, irritated by the oil, was releasing some kind of chemical into the water which encouraged it.

Cleaning rocks was one of the hardest and most labour-intensive operations on the shore. To get the mousse out of crevices and holes, work teams used long suction hoses where they could, and where they could not, large numbers of soldiers were moved in with small cans, and even seashells, to scrape it out into buckets which were then passed back to be emptied into increasing sizes of cans and drums until it could be put into pits or thirty-gallon drums for the vacuum trucks to tackle.

In certain areas heavy construction equipment was put into action. On shingle and cobble beaches, the whole beach was ploughed into furrows, usually at right angles to the sea, or bulldozed down to the surf, on the recommendation of marine biologist advisers, to allow the stones to be cleaned naturally by the sea. In the Ile Grande salt marsh at the north end of the Lannion Peninsula, where more than 7,400 tonnes of oil had buried the area from some inches to, in the channels and creeks, some feet deep, the entire surface had to be scraped up and carted away by truck, after days of high pressure hosing of the mousse into trenches failed to make an appreciable impact. And as the days went by and more and more of the mousse along the coast began to solidify into a state called *sec* in clean-up jargon, front loaders, back hoes and other equipment were brought in to supplement the spade-wielding work teams.

Some areas, such as the islands and reefs offshore, posed particular problems. On the larger ones, teams were able to go out by boat and lay straw, both as an absorbent and in bales as a primitive barrage to keep the mousse at bay. But many were inaccessible, particularly because they were covered with mousse-soaked weed making the rock dangerous to climb, and simply had to be left, creating reservoirs of oil in the centre of otherwise 'cleaned' areas. Offshore, to the constant anger of the clean up teams, large patches of mousse floated tantalisingly, too far out to be reached from shore, and too close in to be tackled by the dispersant boats. On other beaches new and relatively untried methods were attempted, such as the dumping of absorbents like the Michelin rubber 'poudrette' and, on a beach near Trégastel, huge amounts of plaster. Both proved effective but took enormous amounts of labour to collect.

As the major cleaning was completed, fine cleaning techniques were introduced. Rocks, walls, jetties and ramps were sprayed, variously, with steam, hot water, water sprinklers or high pressure hoses, to wash the remaining oil away. On some shores, platoons of soldiers were told to pick up individual oiled rocks, clean them and throw them into the sea. Too often, the sea appeared to return them as black as ever and the soldiers began scratching their initials on the stones and waiting to see if they could find them again the next day. Part of the problem was often caused by the water spraying teams who failed to wash the oil into collection areas but simply allowed it to run down the beach. However this washing was an important finishing process in amenity areas.

As the days of the operation went by, the gradual weathering and ageing of the oil changed the emphasis of the programme. The huge numbers of vehicles which arrived in the area during the first few days were not required as the mousse began to turn from fluid to semi solid form. For example, in Finisterre, while the number of honeywagons remained at or around the ninety mark for most of April and May, the number of tank trucks and vacuum trucks, which could no longer follow the oil into the less accessible sites and which had, in any case, less and less fluid material to carry, fell rapidly, to averages of only thirty or forty a day for the first and twenty to thirty for the second. Dump trucks which had been as many as three hundred on the first day, settled down to around one hundred to one-hundred-and-forty and like all transport which could be used for the *sec* material, were in constant use. Heavy machinery like front loaders and bulldozers remained steady at between thirty and fifty as did trucks, tippers and other wagons in stronger numbers. Indeed, the amount of fluid mousse moved appears to have peaked very early in the operation. Charts produced later for a NOAA/Environmental Protection Agency study showed that the key date for pumped mousse in Finisterre was Wednesday 5 April when 8,000 cubic metres, or around 8,000 tonnes was moved. Mousse shifted by skimmer peaked even earlier, on Friday 31 March at 3,000 cubic metres. By comparison the amounts of bagged material moved did not peak until late in April, when 1,200 cubic metres was moved between Saturday 15 April and Thursday 20 April.

Among those on the beaches, despite the back breaking work and the filth – teams often had to spray each other down with water before they could leave the shore after a shift – there was an enormous willingness to get the job done. Indeed, many of them seemed to consider it a patriotic duty, as if they were making some kind of point to the international oil companies; and the working relationships which developed, particularly between the young soldiers and volunteers and the older civilian operators, enabled a number of problems to be overcome. But over the weeks, inexperience of oil pollution work and a morale badly dented by the apparent powerlessness of the clean up to make anything but slow and steady inroads into the oil threw up inevitable difficulties among the workforce.

The inexperience meant among other things that the teams sometimes added to the pollution themselves, churning up the surfaces of the beaches with their heavy equipment and mixing oil with uncontaminated sand; spilling oil into untouched areas behind the beaches; damaging shore areas by failing to keep to a single access track; and causing erosion problems by stripping contaminated sand from some beaches. And the laboriousness of the work meant that many of the eager young volunteers who had travelled from as far away as Belgium, Holland and Denmark became quickly disillusioned and left; their places were taken by local students from Brest, working through their weekends. The Army took the point and began returning its platoons to a different beach after each rest period so that the men did not become too disheartened. 'What we really need is about a million Chinese,' joked Lieutenant-Colonel Jaunet of the Paris fire brigade at the Ploudalmézeau advance post, early in May.

There were difficulties getting the manning right. The Army in particular had sometimes to learn how to live according to the state of the sea, rather than the barrack room. Dr Molly Spooner, a British marine biologist, visited Poldhu, near Roscoff, where the farmers and their honeywagons had moved 1,500 tonnes of oil in a night early in the spill. There the military had moved in and, reported Spooner, 'At least for some weeks only routine hours were worked: *déjeuné* was a strict break and no one worked after 5 pm no matter how urgent or sensible that they should do so.' The Texas A & M

University team noted that 'Often matching of the tides, times of day and normal workday resulted in extremely short working days for some units . . . for efficiency and economy it is necessary to match men with tools, supplies and equipment in appropriate ratios. For example, if twenty-five men can handle a honeywagon, having one hundred men on the beach with one wagon achieves nothing. Similarly, if twenty-five men can load three honeywagons in series, then having only one wagon in service and having the men wait between loads is wasteful.'

The difficulties extended to commanders as well as men. As the operation went on, the men who had originally been playing central roles at the various levels had to be rested (some had been working from four in the morning until midnight at what even the National Assembly report admitted was a 'frenetic pace'). Even Gérondeau and Di Chiara at Ploudalmézeau had handed over to subordinates. But the rota system was too short, so that each commander hardly had a chance to orientate himself before he was replaced; and, said the Assembly investigation, replacements 'tended to apply forcibly ideas which were not those of his predecessor.' Ploudalmézeau, it was noted, was rarely under the command of one man for more than two or three weeks at a time. Regretting the problems this threw up, the Assembly report conceded that 'an operation of this size had never been attempted before' and said that 'this rotation of top officials under the co-ordination of a member of the Government was improvised' but 'there was no precedent since a situation like this had never happened before.' Later some of the problems were eased when the rotas were revised.

It is difficult now to dispute that a great deal of marvellous effort was put into Operation Teaspoon, with some masterpieces of improvisation; the inquiry report itself picks out the Ministry of Equipment for a mention and says that its work in providing barrages and, on land, bringing up materials and planning the use of personnel and volunteers ensured that it had become 'the backbone of the fight on land'. But if the early days of POLMAR had led to irresolution and confusion through the multiplicity of competences, the now huge size of the operation simply multiplied the possibilities. And there could be no pretending that everything had worked perfectly.

The problem was that as the oil had spread over dozens of kilometres the POLMAR plan had had to be constantly adapted (no one had envisaged, for example, an operation which extended over more than one department while here, until 4 April when Ille-et-Vilaine and Manche ended their POLMAR emergencies, there had been four demanding access to scarce resources either to fight the oil or prepare for its arrival). And the authorities had had to handle liaisons on many levels: with departments, maritime regions, defence zones and all the ministries above them; they had had to negotiate with private industry, with military officials, professional people, volunteers and with foreign advisers. Becam's appointment had been crucial, particularly as a psychological boost for the POLMAR teams, but it could not cut out all the decision-making points, and, concluded the National Assembly report, 'the structure of the organisation tended to become too complex.' The result of that was that the problems faced by each level and each command post were so varied that each spent most of the time trying to solve its own problems in its own way. Some organisations set up their own advance posts to liaise with their own people. The inquiry thought that the trouble was simply that 'the type of structure which established itself had not fundamentally been laid down by anyone' (and it went on to lay out its own ideas for changing all that in future) but it recognised that a lot of the mistakes were made honestly and that 'the complexity of the tasks themselves did not permit a simple organisational scheme'.

In this respect the Army had been a godsend, not only because it had manpower but because it had a proper command system which could also divide its competences cleanly between, for example, men for straightforward cleaning work and those to solve engineering or transport problems. The supply of Army trucks seemed to be limitless. It had a 'natural capacity' to step in where it was needed, said the Assembly investigation, while on the civilian side, local decisions often had to be made in spite of, rather than through, the confused POLMAR system.

A typical point was the use of the young volunteers. Totally unforeseen in the POLMAR plan, they had begun to converge on Brittany in large numbers on the first weekend of Operation Teaspoon: 600 of them came on one train from Belgium alone. And officials, crying about a 'second invasion', worried that they would get in the way, cause disorder, or be the victims of some kind of accident that would cause a new outcry, had had to spend valuable time trying to find a system to handle them, and to find them lodgings. Some clean up co-ordinators were loathe to use them because they were not as disciplined as military personnel. And many were suspected to be loafers who just wanted a free holiday on the State. Initially the local mayors, who had no official tasks under POLMAR, were given the responsibility for them; then a 'volunteer co-ordinator' had emerged in the shape of Lefèvre, the sub-prefect of Lodève; and finally, an official system based on an idea from the Côtes du Nord was put together.

Volunteers would have to report in the first instance to their local office of the Ministry of Youth, Sport and Leisure. Only those over seventeen would be allowed to volunteer and only in parties of thirty with an agreed leader. They would be summoned to work for one week, at 34 francs a day each, and they would have to work when and where directed by local co-ordinators. In Rennes, the regional director of the Ministry, Albert Martin, was forced to set up a special office to keep a check on the exercise and Lefèvre opened an office at Ploudalmézeau to liaise with him. Three reception areas were designated at Tréompan, Santec and Plougasnou with others to be chosen along the coast. And holiday camps were pressed into service for accommodation. But the rearguard system did not stop the 'hands in pocket gangs' from arriving unannounced and roaming the shoreline and local towns; and there was an unfortunate backlash from the community, affecting all those involved in the clean up, when some local café owners and shopkeepers began to put signs in their doors reading 'No clean up volunteers wanted here'.

But there were other, more serious problems. One of the major ones was the handling of the material brought off the shore. In the first few days after the wreck, when few units were at work and the oil was very fluid, this had been easily arranged. Loads were taken to Brest port where barges waited to transfer the mousse to the tank washing station or on to Le Havre for recycling. But as the clean up increased in momentum there was too much material for Brest alone to handle, and in addition to the fluid mousse there was the 'sec' material which the front loaders and work teams had been shovelling off the shoreline. Under the original POLMAR plan, the Ministry of Agriculture and the Bureau of Geological and Mining Research should have identified disposal sites in advance, but no one appeared to have seen the papers and in the event the whole exercise had to be improvised. Recycling plants at Nantes, St Nazaire and La Palice were commandeered as eventual destinations for the more fluid mousse, and large road tankers and railway tankers on the Roscoff line organised as transport. For the solid waste an area inside Brest port for Finisterre, and a small valley near Trégastel for the Côtes du Nord were designated. But interim storage areas to cope with the irregular rhythms of the beach teams, and where mousse and weed could be separated, had to be found in a hurry. In the west particularly, there had been little opportunity for preparation, most were simple pits dug at the top of the beaches or just inland, sometimes lined with polythene, sometimes not. Industry helped by sending some free standing metal tanks, but local POLMAR co-ordinators found their choice of sites for all the interim storage being constantly disputed by technical and professional advisers, local residents, ecologists and the beach teams themselves.

Transport problems also meant that there were constant holdups. One newspaper report said that at one point 569 railway tankers were waiting at or somewhere outside Roscoff, either full and waiting to go out (a small rail bridge on the line meant that no train could weigh more than one thousand tonnes) or standing empty; and as the pace increased two trucks were arriving at Brest every minute, giving place to long queues down the quays waiting to dump *sec* loads. At Roscoff the problems of interim storage and evacuation had become so acute that small tankers and sand ships were chartered to sit in harbour and take the material. It was a decision which caused more difficulties than it solved. Eight ships, totalling 20,000 tons were used at Roscoff and elsewhere and proved extremely expensive. One of the difficulties was that once the material was aboard it was a real task to get it off again. Two of the ships, the *Save* and the *Trieux* sailed from Roscoff to La Rochelle where a specialist company was supposed to treat their cargoes of oiled sand and weed. But when they arrived they were forbidden permission to off load by the port authorities, in case they caused pollution in the bay. The *Save* sailed on to try her luck at St Nazaire while the *Trieux* was caught at the quay by a dock strike. As her cargo started to set solid in her holds, harrassed officials tried to find somewhere to dump it all. They had tried to persuade the master to let them take off part of the cargo by truck to St Nazaire, but the captain, in his last voyage at sea, stubbornly insisted it was everything out, or nothing. At the very same moment another ship, the *Sovereign Star*, was sitting in Brest docks, three quarters full of oiled sand and weed while officials tried to find a place for her cargo.

When another of the ships, the Danish-registered *Henriette Bravo*, foundered in three hundred feet of water in mid-Channel with 2,800 tonnes of oiled weed aboard, the authorities decided ships should not be used again to transport residue. But the decision only added to the question of where to put the oil. One of the major centres turned out to be the site in Brest port. The material was put into the plastic fertiliser bags, of which 200,000 were used in Finisterre alone, and taken to an area where four pits up to an acre in area and ten feet deep had been dug and lined with clay and plastic. The material was to be dropped in and then treated with chalk or quicklime which would, eventually, be used for the foundations of roads and for a new dock which would, with great irony, be used for the repair of supertankers of up to 500,000 tons. But even then there were difficulties: the bags prevented easy treatment of the melange and a crane crew had to be set to work, picking up each individual bag as it arrived and dropping it to make it split open. And it was not a decision which pleased the people of the city: they were to complain about the stench of the oil for weeks to come. In fact the disposal of the recovered material never matched the rhythm of collection. The French were still treating it and finding eventual destinations for it long after Operation Teaspoon was over.

Behind the scenes the French also found themselves in a quiet but rather desperate row over the ban on the use of dispersants along the shoreline. It was one to which the official reports referred only fleetingly later and which French officials made a point of denying or playing down in conducted press visits to see the results of Operation Teaspoon. While the marine biologist advisers and the large majority of the fishermen remained convinced that the chemicals did pose a real threat to the marine environment (and many of the fishermen were not convinced of the explanations put forward for the use of the new generation of 'low toxicity' dispersants offshore) the painstaking physical process of cleaning by the vast numbers of men and machines seemed to others to be progressing far too slowly for comfort. In particular those communities whose livelihood depended on the potential of the summer tourist trade were becoming increasingly edgy and angry as the days went by. Many people there adopted the view that any effect on marine life was of only secondary importance to the possible economic impact on their own and the importance of cleaning the amenity beaches around the resorts in time for the first visitors. Already there had been cancellations because of the publicity of the oil spill and the resort areas were desperate to rescue what they could from what they had assessed as a ruined season. They demanded dispersants be brought in for cleaning. Under pressure the POLMAR leaders gave in but they quickly sought to pass the responsibility back to the communities themselves: dispersants could be used but only at the discretion of the local mayor. A number of resorts went ahead and, said the National Assembly investigation in a passing reference, 'they rarely regretted it'.

The biologists were not so sure. Many of them did not learn of the decision until they saw the chemicals being used on the coast. And even those who held no strong views about dispersants often found extremely worrying incidents. One American NOAA/EPA team found dispersants being sprayed on seawalls at Roscoff, at Santec a week later and between the towns of Ploumanac'h and Perros Guirec at the end of April. 'The operation at Roscoff consisted of men spraying the solution from canisters strapped to their backs prior to washing with water under high pressure. From discussions with the clean-up crew, we learned that a much stronger dispersant had been used earlier but it irritated their eyes and skin and they had switched to a milder one'. And the same team found a large number of dead marine animals on a beach near the village of Corn ar Gazel where the water was a milky colour. 'Although it is possible that the milky water came from offshore where French ships were actively discharging dispersants, these chemicals were also being used by clean-up crews along the shore. In late April a clean-up crew was observed using dispersants and the nearby water turned a milky colour that could still be seen several hours later.' The scientists said that 'Near Santec and Ploumanac'h, firetrucks containing mixtures of water and dispersants were being used. An empty drum at Santec was labelled Treatolite demulsifier, Petrolite Corporation, London.' The use of this particular chemical was not admitted in either of the official reports on cleaning up the spill. In the Côtes du Nord it was reported that the prefect had pursuaded local fishermen to let him use dispersants on the badly oiled port of Lacquémeau.

While the shore spraying never reached enormous proportions, it is still not known how much dispersant was used. Nor has it proved possible to quantify in isolation its impact on the local marine environment.

The entire operation along the Brittany coast line was fraught with difficulties. But one after the other the French sought and, for the most part, found solutions to the problems they faced and as the weeks passed and the clean-up crews worked on, the shoreline came to look clean again. By the first week in May, as coachloads of foreign journalists were being conducted through the area as part of Operation Verité, a public relations exercise by the Tourist Ministry, it looked as if Michel d'Ornano's confident prediction for Operation Teaspoon might yet be right. While it was possible to see a shoreline still badly needing treatment it was also often difficult to tell if particular beaches had once been oiled. POLMAR co-ordinators believed that they had found a powerful ally in the sea itself, which, as well as carrying the oil, had been working since the first day of the spill to disperse it naturally. They saw now that the intense dynamic coastal processes of high wave and tidal energy, combined with the spring storms, had been making an important contribution to their effort. Major difficulties remained for the beach teams in sheltered areas, but the sea had been hard at work scouring exposed rock faces, wave-cut platforms and reefs, and its erosional capacity at this time of the year had enabled it to give some of the fine sand beaches a normal appearance.

During early April major oil accumulations had been broken up and as the weeks went by, the new swash line accumulations began to change into small mousse balls and oiled weed. As the mousse became diffuse and no longer pumpable, it was often a combination of the human clean-up operation and the sea's own action which removed large amounts of difficult material. On one beach between Aber Benoit and l'Aber Wrach,

175 tonnes was reduced to one tonne by 22 April; at Brignogan, 284 tonnes was removed altogether from a beach where it had once been one foot deep; and on the Ile Grande salt marsh 7,400 tonnes of mousse were reduced to only 2,700 tonnes with the help of two hundred men working up to 25 April: the rest was removed by stripping the surface of the marsh with machinery after 30 April.

By the end of May the sense of crisis had gone from the advance posts along the coast and been replaced with an air of relaxed confidence that the end was in sight. Most areas had been officially designated clean and only a few stubborn stretches required the treatment to go on into June. Gradually the number of men and vehicles taking part in the clean-up operation could be reduced and plans made for the entire operation to be stood down. In many areas, local authority teams would take over to continue the fine cleaning through the summer to protect what *Le Monde* called 'the Holy Grail of tourism'. But as the summer sun came out the people of Brittany began to try to assess the damage caused by the *Amoco Cadiz* and whether Operation Teaspoon had actually worked.

10 / A Provisional Balance Sheet

WHEN THE FRENCH LOOKED BACK they found that though the scale of their operation was huge, and they had to move enormous amounts of material, reality had substantially modified the kind of theoretical picture of the clean up which had been put forward in calculations like those of the scientists from Texas A & M University. The astronomic American figures had been suggested as a guide to the potential size of the exercise on the assumption that all the oil which did not evaporate had gone ashore as mousse, and described how it might be transported once it had been cleaned from the coast. But in the event the fate of the oil had been far more varied than their simplistic sketch allowed, and the real figures were more complex.

For in addition to the oil which had found its way visibly ashore, as we have seen, more had been dispersed naturally by the ocean; some was chemically dispersed or treated; some sank to the sea bottom; some was buried in the beach sand; and some remained stranded where it could not be cleaned. The exact figures of how much oil went where are still not accurately known and the estimates may be between twenty and fifty per cent out. Both the French and the American scientists made calculations but could not exactly agree. CNEXO, for the French, reckoned that a little more than 80,000 tonnes of the supertanker's cargo had evaporated into the atmosphere; that 40–50,000 tonnes had probably sunk into the sea bottom; that 64,000 tonnes had gone ashore; and that the 'fraction of oil not included in these estimates' (perhaps 29,000 tonnes) 'is probably buried in the sand of the beaches'.

The Texan team, using locally reported sources, thought 74,000 tonnes had evaporated; 20–26,000 had been chemically dispersed or treated at sea; 25,000 tonnes had been deposited on the sea bed; and 25,000 tonnes had 'disappeared, perhaps under the beach sand'; and that 80,000 tonnes had gone ashore. The essence of these figures was to show just how important the ocean's own power to disperse the oil in its own way had been: though 223,000 tonnes had been spilled and perhaps 74,000 to 80,000 tonnes had evaporated, only between 42 and 53 per cent of what remained had ended visibly on shore (though there was also the large amount believed to be layered into the beach sand). And still it had managed to coat 245 miles of coast. But what proved even more surprising was the amount of oil from all that had gone ashore that the experts calculated Operation Teaspoon had been able to recover. Official figures given to the National Assembly inquiry showed that in the twelve weeks from the beginning of the operation to 26 June, an enormous 206,000 tonnes of material had been cleaned off the shores: more than 17,000 tonnes a week or around twenty-six tonnes per man; an amount almost equal in total to the original weight of the entire *Amoco Cadiz* cargo. But one thing the experts were agreed on was the total amount of oil in all that material: both the French and American scientists concluded that after three months of effort Operation Teaspoon had moved only 25,000 tonnes of oil, or only a sixth of the original cargo allowing for the amount lost to the air by evaporation.

Lucien Laubier of CNEXO said that by mid-summer, the operation had collected 30,000 tonnes of mousse from beaches and rocky areas, but that since the oil content of the mousse was probably as low as thirty per cent, that represented as little as 10,000 tonnes of oil. In addition, he reported, 200,000 cubic metres of the gooey melange of oily sand, weed and beach debris had been moved, but the oil content of this might be as low as two per cent, certainly no more than eleven per cent; say another 15,000 tonnes of oil, giving a total of 25,000 tonnes. The Americans agreed, saying that they thought the total was somewhere between 20,000 and 25,000. The National Assembly inquiry, while remaining in the same area, gave a wider ranging version. It calculated that of the 206,000 metres moved (164,000 of them in Finisterre), 49,000 tonnes had been liquid. Using the Laubier basis of reckoning, the liquid tonnage would represent 14,700 tonnes of oil, and the remainder somewhere between 3,140 and 17,270 tonnes; a total of between 17,840 and 31,970 tonnes of oil recovered.

The experts, and particularly the scientists, were unanimous in their explanations of the low proportion of oil accounted for. The only problem Plan POLMAR had really failed to solve was that it simply did not get started soon enough with its major clean-up. The first twelve days of what the official inquiry called 'irresolution' while the authorities tried to put together a response to the oil, and the decision to delay a full scale attack until after the highest tide of the spring equinox, were thought by the experts to be fatal. If the French had moved sooner and made the effort to start pumping the mousse while it was at its thickest and most fluid in the early days of the spill, and before it had travelled too far, the recovery figures might have been far higher and the amounts lost into the marine environment correspondingly lower. To quote only the Texas A & M University conclusions:

> The longer it takes to clean up the spill, the greater will be the loss of oil to the environment and the more sand, seaweed and detritus there will be with the collected oil . . . Rapid removal of the mousse when it first hits the coast would minimise such losses. Greater dispersion of the spill makes for lighter coatings of more areas. Then removal must deal with larger volumes of lightly contaminated material rather than smaller volumes of more concentrated material. This not only increases the bulk but renders the residue less amenable to treatment and reclamation. It was reported that considerable time was lost in the spill response while policies as to who does what, who pays for what (and how), which methods will be used and what chemicals will be used were being determined. Effective response is possible only when such issues have been decided in advance and the response team adequately informed of the decision.

And the figures did nothing to dissuade the continuing criticism of Plan POLMAR and the overall French strategy to the spill. *Le Monde*, visiting Brittany two months after the grounding of the *Amoco Cadiz*, described it as 'a great, secret and empty dossier' and accused officials of being too optimistic about the success of Operation Teaspoon 'The official version, by a sort of obligation to optimism,' said the newspaper, 'has taken curious liberties with reality' and a great deal of oil remained. The criticisms continued, on a more reserved scale in the Senate inquiry, which appeared just one hundred days after the grounding, and surfaced again in the National Assembly document published some months later.

But at several levels it was agreed that, though the reaction to the spill had not been all it might have been, the disaster was on such a scale that it was doubtful if any country in the world possessed a contingency plan or organisation capable of dealing immediately with something like the *Amoco Cadiz*. Indeed, the horror of the *Amoco Cadiz* caused a number of nations urgently to begin a review of their own arrangements. Britain, one of the few countries to have had a spill approaching the scale of the *Amoco Cadiz* on its shores, set up an investigation into its plans only three weeks after the supertanker grounded, which resulted in a new and specialised Marine Pollution Control Unit, commanded by a retired Rear Admiral, being set up, and government spending on anti-pollution efforts virtually doubled. The British plan, still being criticised within the country as not going nearly far enough, concentrated on increasing its dispersant spraying response, upping the number of vessels on call from ninety to 125, taking on a small fleet of special oil recovery vessels developed by its own Department of

Trade, and providing a retainer for large aircraft equipped for spraying (a response the French had experimented with late in Operation Teaspoon, using a DC4 and two helicopters). The plan also included new telephone and telegraph links across the Channel to the French whose help might be needed in any emergency. The British, even after the lessons of the *Amoco Cadiz*, still saw no case for government involvement in keeping deep-sea tugs standing by at key points; but its plans called for two caches of special equipment, one in Wales and the other in Scotland, which could be used to off-load a vessel of its cargo and bunker fuel and where necessary, inert the cargo tank spaces. A working party investigated how to organise a beach cleaning response on the scale of Operation Teaspoon.

The most detailed calculation of the amount of oil which came ashore from the *Amoco Cadiz* was made by two American scientists from the University of South Carolina, Erich R Gunlach and Miles O Hayes, who were studying beach processes. They estimated that 63,828 tonnes daubed the coast during the first two weeks of the spill and that a month later this had been reduced both by natural dispersion and the energetic clean up operation by eighty-four per cent to only 10,310 tonnes. Ironically, if the basis used by the Texas A & M University team to reach their original theoretical assumptions on mousse generation is applied to this figure, 63,828 tonnes of oil becomes 207,121 tonnes of mousse, which compares very closely with the 206,000 tonnes of materials said to have been moved by Operation Teaspoon. But of course the material moved by Operation Teaspoon was not mousse alone but, as we have seen, contained a very high percentage of extraneous material not allowed for in this reckoning.

For the marine biologists studying the after effects and the ecological impact of the spill, it was not the huge amounts taken off the shores which struck them most forcibly but the massive amounts of oil left somewhere and in some form or another in the marine environment. If only 25,000 tonnes had been cleaned away, then allowing for evaporation, another 124,000 tonnes of oil might remain in the marine milieu.

A major effort to assess the ecological damage from the *Amoco Cadiz* had begun within hours of the wreck. And it was to grow to become one of the largest and most detailed of its kind: even today it is still not complete.

At first, parties of French scientists from a number of organisations moved on to the coast, including those from the Brittany Oceanographic Centre of CNEXO, near Le Conquet (CNEXO is the co-ordinating body for all oceanographic work in France and is under the control of the Ministry for Industry); the University of Western Brittany at Brest; the Biological Station of the Pierre and Marie Curie University of Paris at Roscoff; the ISTPM, the IFP and the National Geographic Institute of France; assisted by students and amateur naturalists from the Society for the Study and Protection of Brittany Nature. At one point nearly one thousand people were involved in the studies. Later the French were supplemented by foreign scientific teams and particularly by the Americans sponsored by the EPA and by NOAA, with whom CNEXO had maintained a bi-lateral study agreement for the previous six years.

CNEXO itself had immediately been contacted by the Ministry of the Environment and ordered to begin a full-scale investigation with promises of credits totalling £357,000 for a first year's work. But it was also to get some outside finance in a unique way. Early in April, Lucien Laubier was approached by Standard Oil, the parent company of Amoco, which made it clear it wanted to fund a long term study into the effects of the oil, to seek what it called 'objective results'.

There was immediate consternation in French Government circles at the idea of carrying out such a sensitive task with money from the company which many felt had been responsible for the pollution in the first place and it was made plain to Laubier the offer was unacceptable. But the deal did go through. Very carefully a special arrangement was devised that would allow CNEXO to take the funds and still preserve the French sense of independence. Under the scheme, Standard Oil would give two million dollars for a three year programme but the money would be paid over to NOAA in America. NOAA, quite separately from its existing agreement with the French, would then form a joint ten-man commission with CNEXO, to be chaired by Laubier, to organise a research programme by both American and French scientists. Laubier would also head up a French committee which, in true bureaucratic style, had been thought necessary to organise the French end of the programme. Though the agreement between the two organisations was never written down, apparently to protect French sensibilities over the money, its last clause agreed that sixty per cent would be spent in France and the other forty per cent in the United States and elsewhere. Eventually nine different organisations were to become involved in research into the effects of the oil and the French received as much as £1.2 million from various sources. Yet in spite of all this research, the complete impact of the oil is still imperfectly understood. The bad weather, with low temperatures and storms had wreaked their own damage and had held back spring with its effects on the reproduction of many species. But the investigations of the scientists, once they began work, produced some of the biggest surprises of the entire *Amoco Cadiz* affair. For they showed that the damage caused by an oil spill is not necessarily in direct proportion to the amount of oil spilled.

There had been some enormous immediate effects in the two or three weeks after the wreck, with millions of small marine animals, particularly those which lived in the intertidal zone or at or below low water mark, and thousands of seabirds, found dead or dying. In the water, while the free swimming fish had been able, for the most part, to flee the oil (only 10,000 were found dead along the entire affected coast), less mobile species at some sites had been all but destroyed, particularly the herbivorous gastropods like limpets and periwinkles that lived by grazing over the weed-covered rocks; and there was massive mortality among the sediment-living animals like the cockles, razorshells, urchins and clams; free moving crustaceans including several species of crab were affected to some extent; and the commercially exploited oyster beds were dramatically damaged with the oysters dead or contaminated.

A research team under Dr Claude Chassé of the University of Western Brittany's Marine Studies Institute checked 160 sites along the coast and found great quantities of sedimentary animals, with those living on rocks 'heavily hit', on the sites checked. Mortality rates varied hugely, from nil to as high as ninety-nine per cent; but an average fifty-four per cent of seven test species were found dead at most sites.

Among the birds, the gulls, gannets, the shags and three of France's less common, the guillemot, the razorbill and the puffin, had all suffered casualties.

The oil, a mixture of the Arabian and Iranian crudes the tanker was carrying, plus some of its bunker fuel, had proved a toxic cocktail, with a high thirty to thirty-five per cent proportion of aromatic components, and up to three per cent benzene. It was reportedly more toxic than the *Torrey Canyon* oil which had had time to shed its more toxic components by evaporation before it reached Brittany. One set of tests found it mid-way in toxicity between Kuwait crude, of the type that had been in the *Torrey Canyon*, and 'Bunker C', a fuel oil notorious for its persistent and significant effects on a number of intertidal animals. The oil contaminated a great deal, particularly the weeds which seemed to hold it like a sponge, by simple coating; but its lightness and low viscosity in the rough seas meant that it dispersed easily throughout the water column, making fine droplets and soluble fractions available to marine life through considerable depths. And at the same time it readily formed mousse which helped keep its toxic components trapped for transport over long distances; the mousse was buried in the beach sands and contaminated the interstitial water; and appeared to have the ability, at least in early life, to revert to pure sheen when stranded and penetrate the sediments. Large amounts mixed with sediment and sank to the sea bottom. At least one study, by six scientists from the Bedford Institute of Oceanography at Dartmouth, Nova Scotia, suggested that the stranded oil became more, rather than less toxic as it weathered, becoming even more lethal than 'Bunker C'.

There were high mortality rates near the wreck and on other parts of the coastline where the oil had been collected by an obstacle, but other shores also suffered. The most dramatic kill of all occurred on a sand flat between the villages of St Efflam and St Michel-en-Grève in Lannion bay, fifty-two miles from Portsall, where on 2 April, as the tide receded, an entire mile-long section of beach was revealed to be covered with million upon million of dead sea creatures or their empty shells.

A swash line of dead heart urchins three hundred yards long and twenty-five yards wide was counted by Gunlach and Hayes of the University of South Carolina and found to contain 120,000 shells; another, two-hundred-and-fifty yards long and six yards wide contained the shells of 45,000 razor clams. At evening tide, at seven o'clock, a five hundred yard section of the beach contained an estimated 1,140,000 dead urchins. 'Therefore several million urchins were present on the surface of the intertidal zone at that time, in addition to hundreds of thousands of dead clams and worms', the scientists reported. Four days later, members of Chassé's team visited the site and calculated that as many as 33 million dead animals of various kinds might have been on an area of sand three miles square. There was no obvious explanation why, at this point so far from the wreck, a kill should have been so cataclysmic, except that it was a large area rich in marine life providing a high number of potential victims; some scientists refused to believe that oil could kill so potently and were forced back into speculation about associated natural storm damage. But deaths of the same kind of animals, on a lesser scale, were noticed all along the affected coast.

But the scientists also discovered something important: that not everything the oil polluted died. The deaths were concentrated in certain species, with an unexpectedly high variability in susceptibility, extreme mortalities in some species and high survival in others, for which there were no immediate explanations. The Bedford Institute team suggested species living in open water were more susceptible than those which could burrow in the sediment. And they also suggested that some species actually had the ability somehow to detoxify themselves. It was an improperly understood process, they wrote in a research paper later, but 'it is interesting to speculate that the survival of mullet in the awesomely oiled salt marsh channels near

Ile Grande may in part have been due to their ability . . . to successfully reduce or eliminate petroleum hydrocarbons from their tissues.' Other scientists took a different view: the great mortalities close to the wreck could be attributed directly to the aromatics, while the lesser deaths, usually further away, showed that some of the toxic elements had had a chance to evaporate. The first effects of the oil were probably narcotic so that the animals either lost their hold on the rocks or came up out of their burrows and were tossed ashore by the surf to die at high tide mark along the beach. The difference between the survival of the species might be that the sensitive creatures, actually those living below low tide mark, never recovered before they died, while the hardier creatures, those which lived higher on the shore and had known chronic pollution, recovered from their comatose state once the oil had been carried away or stranded on the beach at high water mark.

A prime example of survivors among the flora were the various seaweeds along the coast: despite heavy coatings of oil very little of the weed actually died. In some cases observations showed that its edges were damaged, that it became more brittle, that its holdfast was less firmly attached to the rocks, or that its growth was not quite so fertile, but the biggest losses of weed were from natural storm damage or where it had been cut away by clean up teams. Further out to sea this did not take place and the weed remained whole and healthy, probably helped by the mucus in its membranes, even if it also held oil which might contaminate other species.

A Dutch team examined seven species among the estimated 245 square miles of weed on the coast, which had been seen to suffer from the *Torrey Canyon* oil and reported 'no significant mortality'. In particular those species harvested by the seaweed fishermen had not died though they were often oiled. Some scientists postulated that this was due to the relatively small amounts of dispersants used under POLMAR (as opposed to the 10,000 tons used by the British on 14,000 tons of oil on ninety-three miles of their shores from the *Torrey Canyon* in 1967) and which were known to have adverse effects on weeds. Somewhere around 1,500 tonnes of concentrated dilutable third generation dispersant, and hydrocarbon solvent-based versions, including BP1100X, BP1100WD, Fina OSR5 and 7, Fina OSR2 and Finasol CC were used on the spill, enough, in ideal conditions, to disperse about 12,000 tonnes of oil. But the conditions were far from ideal and few of the boats, except the British, had proper equipment.

The NOAA/EPA report of the clean up concluded that: 'The effectiveness of dispersants as a control tool was limited by the broad distribution of oil along the coast, the patchiness of the oil in windrows and the ability of a ship to cover only a limited area even during the better weather conditions. It is believed that considerably more oil was dispersed in the water column as a result of the high seas and the action of the seas with rocky areas than was removed with dispersants.'

The difficulty of assessing the impact of the oil was shown clearly by the case of the birds. Both French and British ornithologists investigated, finding that up to the end of March only 4,572 oiled birds were collected from the beaches, with the auk family being the most abundant casualties: 1,391 puffins, 978 razorbills and 731 guillemots. A total of thirty-three species were affected including 126 rare divers. But there was the question of whether this represented all the birds which had died. Jean-Yves Monnat of the Marine Studies Institute, working independently of the British ornithologists from the Royal Society for the Protection of Birds, extrapolated the figures to speculate that as many as 15,000 to 20,000 birds might actually have been killed. The British, led by Peter Hope-Jones, were more conservative; though they mentioned no speculative figure in a paper they wrote later with Monnat, a drift experiment they had carried out by dropping the corpses of gulls offshore turned up three (6 per cent) out of a batch of forty-eight dropped thirty kilometres out and twenty-nine (thirty per cent) out of two batches dropped fifteen and 7.5 kilometres out. All came ashore within seventy kilometres of Finisterre coastline. A total of 1,035 oiled birds had been brought from that section of shore between 30 March and 25 May, so if the thirty per cent finding and reporting rate in the experiment was typical at least 3,450 more birds should have died.

In fact the paper accepted that the real mortality rate might actually be higher than this. Extrapolating the full results of the experiment, 21.5 per cent of the 144 birds dropped came ashore; if this were true for the whole of the affected coastline, on the basis of the March collection figure alone of 4,572, the total kill might equal around 22,000 birds.

Many other birds, like the waders, had simply disappeared from their normal habitats; and some ecological detective work was necessary to discover that not all that had happened could be directly attributed to the oil. For example, puffins appeared to represent a large proportion of the kill and yet these birds were not normally seen in large numbers in Brittany that early in the spring. In addition many corpses were found to be emaciated and with flight feathers missing; many bodies were badly decomposed suggesting they had been dead for some time. On investigation it was found that unusually large numbers of puffins had been seen in the Bay of Biscay, including British-ringed birds, from the end of January. They appeared to have been migrating north and been caught in what the ornithologists describe as a 'weather wreck', a bad period of weather which throws the birds ashore in large numbers and unexpected places. The storms in Biscay had led the birds to flee north faster than usual, arriving in Brittany debilitated and exhausted, with many of them in moult and unable to fly, just in time to meet the oil pouring out of the wreck. Like the other auks they are also swimming and diving birds who would want to spend time in the water before going ashore to nest or rest.

Some of the missing birds had not died but had sought other habitats: some were not true marine birds and had simply moved inland; others, like the waders, had had to seek other shores after the massive mortalities of marine invertebrates had cut off their main source of food.

One of the serious points about the bird deaths was that many of those which died were not from local colonies but migrating through the Atlantic approaches to breeding grounds elsewhere, whose populations might be affected as a result. In particular sixty-six great northern divers died which perhaps came from a small breeding colony of three hundred pairs in Iceland. The ornithologists' paper said 'The *Amoco Cadiz* incident once again emphasised that there are large numbers of birds vulnerable to oil spills in the Channel and its approaches in the early spring. The impact area extends to breeding populations far outside the immediate area; to seabird stations of the Irish Sea and northern Scotland and great northern divers which may breed as far afield as Greenland and Canada.'

But the deaths also meant a heartbreaking job for the members of the French League for the Protection of Birds who administered the Sept Iles bird sanctuary off Trégastel. The seven islands and rock outcroppings, sixty-two miles from the wreck site, are under government control and managed by the League; all human access is banned except for the Ile Aux Moines which can be visited with special permission. But no such permission was needed by the oil which swept all round their shores. The islands had been badly affected by the *Torrey Canyon* oil which had much reduced the size of the breeding colonies there. For example a colony of puffins 2,000 strong was decimated and had slowly climbed back to 800 in the intervening years; 250 razorbills and 400 guillemots had been reduced to 90 and 150 breeding pairs respectively. Painstakingly, the administrators, with the help of a grant from the Berger Foundation with which they built artificial nesting sites, had been encouraging the birds to return. Now they would have to go out in their outboard-powered dinghies to the islands again and make a count; it seemed as if the struggle would have to begin all over again.

But dramatic as the figures for the bird deaths seemed, ornithologists were surprised – by comparison with the amount of oil spilled, they were low. The possible 20,000 deaths represented only 0.1 bird per ton of oil; by comparison the 117,000 tons spilled from the *Torrey Canyon* had killed 30,000 birds (0.26 per ton) and a small 6–700 tons from a tanker called the *Hamilton Trader* had killed 6,000 (9.2 per ton). Indeed in one black accident one ton of oil spilled in the Firth of Forth in Scotland during 1978 was to kill 800 birds. Monnat offered three explanations: the large numbers of birds temporarily flightless which did not go into the water; the absence of many species of birds which were still migrating; and the disappearance of many birds inland. The lesson seemed to be that it was where and when oil spilled, as well as its quantity, which determined its lethality.

As important as the immediate lethal effects of the oil, was the rapidity with which things began to change. Both the energetic human clean-up, the much-aligned Operation Teaspoon, and the natural power of the ocean were combining to remove large quantities of oil. After the early localised heavy kills the high mortality rates started to decline quickly and, except in areas where there were special circumstances, such as the sheltered muddy-bottomed *abers* and the bay of Morlaix, the hydrocarbon levels measured in the water began to fall. One by one, species which had fled began to return to their habitats, young fish were seen swimming and darting quite close to the wreck; the weeds and plants began to grow, rock samphire was seen sprouting defiantly through an inch of stranded mousse; and the marine animals started to re-establish their colonies; groups of limpets were seen grazing over lightly-oiled rocks, playing their own small part in the cleaning up.

The natural cleaning power of the sea in particular was not to be underestimated: even without the turbulence of the recent bad weather, the coastal waters, which often fell to depths of 180 to 360 feet within four miles of shore, were subject to currents running parallel to the land of three to four knots (one channel north of Roscoff moved water at two yards a second) and there were north-south tidal currents in the bays and estuaries of two knots; those, coupled with tides that rose and fell twenty-four to thirty-six feet had a substantial effect on the oil.

At some of the exposed high energy sites scientists were reporting hydrocarbon readings by mid-May which appeared not much more than normal background levels. The levels remained high where the oil clung on: in the sheltered inshore areas, the narrows, the creeks, around the islets and in the coves where Operation Teaspoon was having laboriously to clear away. But the French scientists, at least, were confident that even there, with the exception of the *abers*, levels would fall to acceptable levels within months. By the first week in June they felt confident enough of their initial inquiries to present them to a one day conference of the International Council for the Exploration of the Sea at Brest. The overriding message from many of the papers was summed up by Claude Chassé, writing some months later: 'By comparison with previous pollution accidents and taking into account the enormous quantity of oil spilled at Portsall, the provisional balance sheet indicates rather less damage to the marine environment than might have been expected.' Chassé put down the cause to the combination of human clean-up and natural dispersion of the oil, but above all to the limited use of dispersants.

One particular indicator which encouraged Chassé and British biologists who visited the scene was the absence of 'greening' in the rocky areas. When oil spills on a coastline and kills large weeds and the grazing herbivores, its path is usually followed by the flourishing of a 'green flush' of certain small algae on the rocks. This flourishing, in the absence of other organisms, is taken by many scientists as a sign of gross imbalance in the ecosystem and was predominant on the Cornish coast of England after the *Torrey Canyon* spill.

In the years which follow, the normal 'mosaic' of flora and fauna is disturbed as the coast swings between dominant species. The green algae is followed by forests of dense brown weed, flourishing in the absence of the herbivores. In turn the herbivores, encouraged by this large food source return in armies, grazing across the rocks in huge moving 'fronts' until almost all the weed is gone and they die once again, this time of starvation, to be replaced by the barnacles. According to Dr Alan J. Southward of the Marine Biological Association of the United Kingdom, in Plymouth, it took at least ten years of this kind of pendulum-like progression before the Cornish coast returned to a normal distribution 'mosaic' after the *Torrey Canyon*, with unknown repercussions for other species in the interim. Though some such 'greening' occurred in Brittany showing a similar process starting, it was on a far smaller scale, which led scientists to assume that the imbalance in the ecosystem was perhaps less severe. And by July Southward felt able to make a general forecast about the whole affair: 'Even where the shores in Brittany are still oily, it can be predicted that, if the ban on dispersants continues, a fairly complete recovery is possible in three years.' It was a prediction shared by many of the scientists who studied the *Amoco Cadiz* spill.

The relative rapidity of the short term recovery of the coast was particularly noticed by those who made their living from the sea. In the case of the

A honey wagon loads up oil which has been scraped into pits on a badly oiled beach.

Top The oil arrives at an interim storage site to be loaded into road tankers for transport to Brest. *Bottom* The wreck of the *Argo Merchant* off Nantucket, 15 December 1976.

fishermen, when the oil came, some of their boats which could fish in deep water had still been able to get away and bring in normal catches. But the majority of the fleet were small, inshore vessels adapted to several types of fishing according to the season of the year, and it had been these vessels which had been trapped in port. Already in harbour for some weeks through bad weather, they found their stay being extended as the oil covered or threatened their normal fishing grounds. Some boats, 'courageously' said the National Assembly investigation, did try to fish, sometimes by changing their catch or by seeking new grounds in unaffected waters. But alternative catches were not ones normally fished for the time of year and were small in number or immature; and boats often had to travel exhausting distances to find new grounds. Yet by late April the fish were beginning to return to local waters, and by June, according to the National Assembly report, things were almost back to normal. And after three months inactivity, the 1,310 men and 681 boats which fished the Brest and Morlaix districts from a sprinkling of little ports were able slowly to start work again. Catches remained reduced in the Brest district which was nearest the wreck and some individual species appeared heavily hit: mullet in particular. The annual catch for this fish, normally caught in spring, was practically nil. But according to the official statistics, things improved quickly in the Morlaix quarter and catches in general by July and August were better than those for 1977.

For those fishermen who caught crustaceans, lobsters, crabs, crayfish, shrimps, the picture seemed the same. Many of these animals, particularly the spider crab, had been about to move from deeper water into the coastal shallows to breed when the oil struck and so were not as numerous in the most affected areas as they might have been later in the year. The National Assembly report said that there had been 'no apparent mortality' among the commercial species. A calculation of the living stocks between Porspoder and the Ile de Batz showed that contaminated animals probably did not exceed five per cent; and some untainted animals were even caught at Portsall in May. Such figures were a relief for the north Brittany fishermen who brought in 7,000 tonnes of crustaceans a year, or a third of the national production (4,300 tonnes from the Brest and Morlaix districts alone and fifty per cent of all the lobsters in France came from Finisterre). The contaminated crustaceans appeared to come from well defined areas, which ISTPM scientists were able to communicate to the fishermen. Nevertheless the restrictions on harvesting areas meant a drop in catch for the season.

The seaweed fishermen who had wondered whether their harvesting season, due to begin on 15 April, would get under way, found the oil had cleared sufficiently to allow a start just four weeks late. Five different weeds were harvested along the coast, mainly between Le Conquet and Brignogan, producing between seventy-four and ninety-three per cent of the national total depending on the species. Large areas of weed remained coated with oil and of the two main families collected one was affected more than the other. The *Laminaria*, collected from subtidal waters up to seventy-five feet deep by 200 fishermen using 150 special cutting boats, and then laid in the meadows behind the beaches to dry, produced between 4,000 and 6,000 tonnes a year for such products as bath liquids and shampoos and a jelly used in the cosmetic industry. Because the factory processes used to treat it eliminated gross hydrocarbons its harvest was less affected. But the *Fucus*, intertidal weeds picked by hand from the shore at low tide, had been badly hit. Badly oiled because of where it grew, it had also been cleared in large quantities by the clean up teams. And its harvesting was generally useless: the 6,000 to 7,000 dry tonnes usually produced went for flours for animal feeds and to produce additives for over 300 foods including beer and jam; and unlike the *Laminaria*, the processes used to treat it could not get rid of the hydrocarbons. The fishermen were forced to make up the shortage by tackling new fields, particularly around the Ile de Molène.

But in the midst of this scene of general recovery there was one overshadowing disaster area: in the *abers* and the bay and estuary of Morlaix which were the centre of the oyster industry. The industry, which produced 10–11,000 tonnes of the molluscs a year, or around ten to twelve per cent of national production, was, to all intents and purposes, wiped out.

Ironically, it had just recovered from another disaster, this time natural: a parasite which from 1968 had devastated the breeding of the native flat or plate oyster, *Ostrea edulis*. The industry had been saved only by the introduction of the hardier imported strain of cupped oyster, *Crassostrea gigas*. All the oyster bed sites were relatively near to the wreck and open to the west from where the wind had first brought the oil. The oil had come into the *abers* following the failure of the barrages, and insinuated its way into the estuary at Morlaix and La Penzé, where it had been trapped. It arrived at a density, said the National Assembly report, close to that of seawater that was almost unstoppable, and in an emulsion that made it easy for the oysters and the sediments to accept. Frantic efforts had been made by many breeders to remove their oysters from their racks and transfer them to areas of clean water elsewhere, particularly in southern Brittany. This had been achieved on a larger scale at Morlaix, where people had had more time to react, but inevitably thousands of tonnes of oysters remained. Of those not transferred, fifty per cent of those in the Aber Benoit, most of them the expensive flat oyster, died; the total was slightly less in l'Aber Wrach and at Morlaix it was twenty per cent. But those which remained were unsaleable because of the level of hydrocarbon contamination. The sheltered waters did not lose their oil quickly and the black mud of the oyster beds acted as a geochemical trap; though the ground water prevented the oil from soaking in directly, the thousands of holes and galleries made by marine creatures gave an efficient means of entry, often centimetres down. Oysters have the ability to take in and pump out large quantities of water and to cleanse themselves of impurities but the contamination levels here would be maintained by hydrocarbons leaching from the mud. And it was clear the levels would remain high for a considerable time.

In the *abers* in April levels were 150–250 parts per million compared with an open waters level in clean seas which might be as low as 0.020 ppm. in May they rose to 450–650 ppm and then stabilised at 100–200 ppm. In Morlaix and La Penzé, the April levels were 150–200 ppm, which dropped to 100–150 ppm in May and held. Checks on all sites in July showed levels actually rising, contrary to scientists' expectations, with a slight fall in August. And an experiment in introducing oysters from outside waters illustrated just how potent the oil still was: in two months the levels in the oysters went up from 40 to 100 ppm; 'this shows the scale of the disaster,' said the National Assembly report. By comparison, on the other side of Brittany, oysters taken from the *abers* when contaminated and put into the unoiled river estuary at La Trinité, fell from 250 ppm to 60–70 ppm in twenty days. Since it was not possible to transfer all the remaining oysters to healthy waters, the breeders and the French authorities, after the July readings, came to a grim decision: all stocks in the *abers* would be destroyed together with all those over 70 grammes in the Morlaix area.

The seriousness of the situation in the oyster industry emphasised the point that while the coast was being returned to its habitual aesthetic beauty, if not ecological equilibrium, the economic impact on activities either closed down or interrupted by the black tide from the *Amoco Cadiz* was bound to be considerable.

Calculations are still being made, based on hundreds of claims, dossiers for compensation, and levels of social security payments, but it is clear that the damage, both in monetary loss and temporary and sometimes permanent unemployment, has been enormous.

In October 1978 the Finisterre Departmental Development and Management Committee looked back to try to gauge the direct impact on the fishing industries. It examined the Brest and Morlaix districts and compared catches for the six months of March to August 1977 to those for the same months of 1978. In fish the Brest district was forty-three per cent down in tonnage, with ports near the wreck like Portsall and Kerlouan particularly hit; and the value of the catch was down thirty per cent. In Morlaix district the tonnage was identical to that of 1977: after an important deficit in May, the catches for June, July and August were higher than for 1977, and the value of the catch was up twenty-two per cent because of the late good results. No explanation was given but officials suggest skippers were putting in more effort to wipe out the early season deficit. Crustaceans in Brest district were down thirty-two per cent on catch and twenty-six per cent on value; in Morlaix, despite catches in July and August above 1977, they were down 15 per cent on tonnage and six per cent in value. Seaweed in the Brest quarter actually went up in tonnage by fifty per cent and up sixteen per cent in value; and in Morlaix it went up 67 per cent and up 27 per cent in value. 'The results for weed seem paradoxical but in fact the rise is due in great part to the catches from Ile Molène and to the fact that the 1978 tonnage is expressed entirely in "green" (wet) harvest whereas the 1977 tonnage was partly "green" and partly "dried",' says the report.

But the biggest deficit came when it looked at the oyster industry: lumping oysters and the small coquille catches in the areas together, the Brest quarter, which includes the *abers*, was down 80 per cent on catch and 66 per cent down on value; in Morlaix the catch was down 62 per cent with the figure for June, July and August representing conquilles only, and 62 per cent down in value.

Significant as these kinds of figures may be they do not tell the full story. For in addition to the actual damage attributed to the oil, the Bretons also had to cope with what they called 'consumer psychosis'. So extensive had been the publicity over the spill that consumers had come to believe that anything and everything from Brittany was covered in oil. Not only would they not buy fish from the waters off the affected coast, they began to turn against fish from any part of Brittany. They refused even to eat lobsters which had been fished off the coast of Africa because they had been caught by Breton boats. Housewives stopped buying Brittany vegetables because they believed oil might have been sprayed on to them, even when they came from departments far from the sea. And above all, holidaymakers stayed away from the region, not only from Finisterre and the Côtes du Nord, but from all departments, even ringing to cancel holidays in the south and giving oil as the excuse. The situation became so acute that Marc Becam was forced to make radio and television appeals for commonsense.

When asked what people could do to help over the oil, one of his stock replies became 'Buy Breton'.

Indeed, the destruction of the oyster stocks had less to do with the potential physiological condition of the animals than the breeders' need to re-establish customer confidence: Brittany oyster, once a phrase that would guarantee a sale, now became a commercial albatross. The National Assembly inquiry report wondered whether some of the poorer breeders would be able to start up again, 'In the *abers* the situation is dramatically simple: because all stocks have been destroyed the only question now is whether it will be possible to start up oyster growing again after several years.' Could breeders, with no money to keep staff on, no stock to re-seed the beds and no oysters coming out for at least eighteen months survive the financial strain. One dealer interviewed by American biologists at St Pabu on the Aber Benoit said he had laid off 119 out of 125 employees. Long term plans for the *abers* include special vessels to dredge the oil retaining pits and hollows in the muddy bottoms; and the breaking up of the sediments to release oil and the re-making of the oyster beds. In December 1978 tentative trials were being carried out using Irish oysters.

The 'consumer psychosis' particularly affected the holiday industry and was noticeable for the numbers of foreign tourists who failed to arrive, especially the Germans. These were important visitors to Brittany. The region had built itself up to become the second most popular tourist region in France, after the Var, using only the attraction of its coastal strip. But the climate meant the season was short.

The French came mainly between 10 July and 20 August, while the foreign visitors, mainly German, British and Swiss were happy to come in June and September and the 1,063,316 of them who had come in 1977 had been very welcome for extending the summer. But in 1978 it was a different story. The global figures produced by the Development and Management Committee showed the number of tourists in Finisterre, for example, up to 50 per cent down for the months up to June; 36 per cent down for July and 15 to 20 per cent down for August, in all a drop of about 7 million tourist days as compared with 1977. Hotels and camping were badly hit. For hotels, Easter, April and May were 30 per cent of 1977; June 50–55 per cent; July 60–70 per cent in the first two weeks and 80–90 per cent in the second two; and August 90–98 per cent. Camping had a 'catastrophic' June, a bad beginning to July with the second half a little better. In August camp sites in south Finisterre were full while those in the north were not, but even in the south there was a drop in customers for those sites on the coast. 'In total, the 1978 season is the worst that the Finisterre hotel industry has ever known' the committee reported. Even second homes and rented houses and cottages, one of the areas biggest sources of holiday beds were hit: June was 60 per cent of 1977; July down 20 per cent; August, full; and September down 50 per cent, with appreciably fewer foreign visitors in June and July. The committee said in general that 'the 1978 season was very short with the real season only lasting for a month'. The impact was the more invidious in its effect on family incomes: bedroom and kitchen staff laid off were often the wives or relatives of fishermen who were already hit.

The Government had tried to establish immediate compensation schemes for the obvious victims of the spill: the first meeting of ministers had been at the Hôtel Matignon on the morning after the wreck. Fishing boat skippers, for example, were told they could have £148 per man for every fifteen days the man was kept in port and up to seventy per cent of the cost of preparing for sea again; this was replaced from May by a general scheme to hear claims for standing charges such as boat mortgages, insurance, tax and social security costs. Oyster and shellfish growers were compensated for the cost of moving stock, the oyster breeders getting 47p per kilo for cupped oysters destroyed and £1.78 per kilo for flat oysters, plus help with staff laid off up to £296 per man. The bill ran into millions of pounds by the end of the year.

But many dossiers were not covered by these specific arrangements and are still to be sorted out; many businesses depending on fishing for their income were also hit including boat repairing, equipment makers, canneries and transport firms; as were other small firms making use of the sea, such as marine sand and gravel companies. The compensation of the tourist industry was to wait unil the end of the year so as to see the full effect of the season, but while the official scheme covered the affected coast where trade was down an average of fifty per cent, it did not cover the southern areas hit by 'consumer psychosis' which were down thirty per cent.

Other things were hit. The Regional Tourism Committee, explaining that 'the season of 1978 will be the worst we have ever known' went on to say that it would have 'repercussions on all branches of activity relevant to tourism . . . but also on a multitude of jobs which in the course of the summer see their turnover progressing in a spectacular fashion, such as grocers, garage and petrol station owners, tobacconists and so on'. Special offices were set up to deal with the compensation schemes set up by the Government and the claims against short term funds released by the various ministries. But thousands of people suffered some kind of impact on their lives because of the oil and the backwash of economic effects through the community; hundreds were laid off or sacked; and as all the various activities and firms slowly recover from the catastrophe (the oyster industry, for example, late in 1978 was cautiously optimistic that some of the small oysters remaining in parts of the bay of Morlaix might purify themselves and grow large enough to have some kind of harvest in 1979 and 1980 however weak; and tourism was being helped by a massive Government publicity campaign) it is clear that some people may have joined France's growing number of long term unemployed; and that some of the older people put out of their jobs may never work again.

Up to December 1978 a total of £3.47 million in compensation had been paid out to people and firms in activities immediately connected with the sea.

Yet the impact might have been far worse had it not been for the wave of national and international sympathy for the plight of the Bretons. Stirred by the publicity given to the disasters, thousands of individuals and organisations from all over the world sent cash donations large and small to the region. In France itself, the solidarity was encouraged by Marc Becam, who appealed to towns and cities across the country to 'adopt' villages and towns on the blackened coast, and the country rallied to his call: Paris, the capital, with the enthusiastic support of its mayor Jacques Chirac, adopted Roscoff and the Ile de Batz; overseas territories with French links were moved to help; and those towns which were 'twinned' with others in foreign countries were sent donations by their 'twins'. Even in Fiji local people clubbed together to send a contribution. Some of the money was sent to individual communities along the coast, some was sent for general relief; some came from the purses of widows and pensioners in a few francs and centimes; some from large scale fund raising activities, sometimes by the Press and other media, and from public collections. By November 1978, over £1 million was collected in the coffers of the two departments. At the end of October, Finisterre alone had received a total of £773,000 and a special Departmental Intervention Fund for Damage by the Black Tide, consisting of six regional councillors and five local mayors, had to be set up to decide how to dispense the 'products of public generosity'.

By the end of the same month £206,000 had been given away, including £160,000 to those who had been hit the hardest, the fishermen, the oyster growers, the shellfishermen and others directly affected by the oil, on the basis of the compensation claims they had submitted; £24,400 went in advances to firms to help them keep on staff who might otherwise have been laid off; £21,400 was sent to local councils because of donors' wishes. Another regional council fund disbursed £178,500 including £14,800 in advances to commercial companies for loss of business and £22,600 to the local tourism committees for the Brest and Morlaix tourist offices. Even after the payouts, £388,690 still lay in the departments' accounts and £139,285 in the Intervention Fund books and the National Assembly report was criticising local officials for not handing the cash over, given the urgency of the situation and the fact that 'the employment of these financial means would allow the resolution of urgent problems produced by the *marée noire.*' One of the problems was that £279,600 of the £773,000 received by Finisterre had been given directly to local communities and was not available for general dispensation: and there was great discrepancy between the amounts donated, with one community getting only £119 and another £119,000. In the Côtes du Nord, £315,000 had been received by the end of October, and a special commission gave £11,500 for urgent need, including £8,300 for social aid in the disaster communities and £3,200 for individual hardship cases; £25,000 was given to companies in difficulty; £48,800 to local councils; £3,700 to the local fishermen's committee at Lannion; and £5,900 in legal fees. In November the commission met again and handed out another £119,000; by the end of that month local communities had received £53,600.

In many cases the donations had been given for a specific purpose when they were more urgently needed elsewhere and the National Assembly inquiry called on the Government to make it easier for the monies left in the donation accounts to be used with more freedom. But for the most part local mayors were able to use funds given to their communities as they wanted. In Portsall, Alphonse Arzel and his colleagues found themselves able to pay out thousands of francs to local people in difficulties, particularly those with young families. One third of the money was paid over immediately and two thirds kept back for longer term contingencies; 125 families benefited from a £45,000 fund.

There was also European solidarity from within the Common Market; the European Community donated £333,000, paid into the Ministry of the Environment's Anti-Pollution Intervention Fund, of which £200,000 was to go to local councils to help pay for their share of the appalling costs of cleaning up after the tanker wreck. For compared to other incidents along the coast the financial burden of the *Amoco Cadiz* was spectacular and not all the bills could immediately be reimbursed by central government. The *Olympic Bravery* had cost £547,000 of which £517,000 had eventually been recouped through the TOVALOP scheme; in the case of the *Böehlen*, the cost was £18 million because of the pumping operation undertaken by the French, of which only £738,000 ever came back from the owners. By contrast, the first figure worked out for the *Amoco Cadiz* in early summer was an stronomic £49 million and probably as much as £55 million by the time the figures were looked at in more detail. This was only the figure for the clean up and did not include any compensation for the economic consequences.

The bill, for personnel, services of private companies, military and naval operations, treatment of recovered melange from the shore; cost of planes, helicopters and ships and the purchase of equipment (including some expensive Norwegian skimmers) was broken down according to who spent the money and came out, provisionally, as: Prefecture of Finisterre £16.69m; Côtes du Nord £7.26m; Ille-et-Vilaine £0.20m; Manche £0.095m; Préfet Maritime, Brest £3.95m; Préfet Maritime, Cherbourg £0.059m; Ministry of Interior £1.23m; Ministry of Defence £18.45m; Ministry of Transport £0.13m; Ministry of Environment £0.83m; Marché Langeberg (for the skimmers) £0.46m. But by the time these figures were given in the National Assembly investigation report at the end of 1978, not all the bills had been received, for the costs also included many other items such as accommodating the volunteers, paying for damage caused by the clean up teams, petrol for clean up vehicles, the cost of the operation to save oiled seabirds, and the accommodation of recruited civilian workers. And the National Assembly report made it plain that government aid and the millions of francs pouring in from donations ought not to blur the fact that the damage was caused by the *Amoco Cadiz* and its cargo; and, said the report, that was where the eventual cost should lie: with the tanker's owners.

It is clear that the economic consequences of the *Amoco Cadiz* disaster will be with Brittany for some time to come. But for once Paris has been ready to support its outlying province: when Jean Rouyer, president of the Regional Chamber of Commerce and Industry had an interview with André-François Ponçet, secretary-general of the President's office to ask about compensation, Ponçet was quite clear: the payments would be kept up 'four, five, six years if necessary'.

Only the continuing studies planned by the scientists for the coming years will finally put the ecological damage into its full perspective. But the consensus does remain among many of the scientists who worked along the coast in the summer of 1978 that the immediate observable effects of the *Amoco Cadiz* did not appear cataclysmic when set against the enormous amount of oil spilled. Many were impressed and sobered by what had happened, with Lucien Laubier introducing the collected papers from the June conference in Brest by calling it 'one of the greatest ecological disasters in marine history'; Southward, in England, wrote about the 'appalling biological effects of the disaster' and Richard A. Frank, administrator of

NOAA, said the whole affair was 'devastating'. But they also recalled the intermixing ability of the rich marine life throughout the Western Approaches, the natural cleansing power of the seas in the area, and the fact that they had been able to have a beneficial influence on the way the human clean-up operation had been conducted, particularly in limiting the application of chemical dispersants. And the general pervading attitude in scientific circles seemed to be expressed by the British biologist Dr Molly Spooner in her introduction to a digest of papers on the spill in the influential *Marine Pollution Bulletin*: 'While sympathising with the disturbance caused,' she wrote, 'it should be pointed out that fortunately the worst fears early expressed by uninformed sources have, as usual, been by no means realised.' And the French were very pleased with the part played by their clean-up; the National Assembly report said it had made an 'irreplaceable contribution' to limiting damage; and Alphonse Arzel, the mayor of Portsall, was able to sum up for the local people along the coast when he marvelled 'When I first saw it all, I thought it would take a year to clear away; but we were clean for the summer!' And indeed it was possible to be optimistic: most commercially exploited species seemed unaffected; most of the larger sea creatures had survived; most mortalities had been confined in a localised, selective and partial (if sometimes heavy) way to the small marine animals in the subtidal and intertidal zones and the birds (always, in any case, the most visible victims of an oil spill); and most plant life had escaped unscathed. No species appeared to have been completely wiped out; no rare species lost; and the diversity of species might only be temporarily interrupted. Indeed, on the best estimates most of the coastline would be restored in balance and numbers within three years: substantial sums had been set aside for the re-seeding of at least the commercial open water species to ensure that this would be so. Only the oyster culture gave really deep cause for concern.

Claude Chassé, in a paper written late in 1978, estimated, in an attempt to illustrate the moderate effects of the spill that, in all, thirty per cent of the fauna and five per cent of the flora on the affected coasts had died. But while the figure for plant life was perhaps not significant his estimate still showed that almost one third of all the living creatures along the oil daubed section of the north Brittany coast were killed. And the question mark still hanging over the region is whether it is possible to be reassured about the long term effects of the oil.

For only the oil removed by the clean up operation and that which evaporated into the air can truly said to have left the marine environment. Though the natural processes of biological and photo-chemical degradation will have worked substantially on what remained (for instance, that on the rocks would be broken down by ultra-violet light; herbivores grazing over weed and algae would swallow the residue and pass it in their faecal pellets to be dealt with by the endless bacteria and micro-organisms) all is not over in Brittany. In the *abers*, the seriousness of the position is acknowledged: the ability of the muddy bottoms, their waters relatively unmoved compared with the open coast, to soak up oil may mean substantial hydrocarbon levels for many years (certainly high enough to have repercussions on the cultivation of oysters) and the estimates range from four to ten years before they will be clean, depending on the success of operations, particularly the dredging of the debris-lined furrows in the bottom where oil has gathered.

But there is still oil elsewhere on the coast and how much and what effects it might have is imperfectly understood and under urgent investigation. Large quantities of oil are known to be buried up to forty inches deep in the beach sands, layers which will be exposed once again during the storms of the coming winters; and which some research insists will re-emerge more toxic than when it was covered. Not only did the oil go into the sand in an early form (when the mousse gave readings of hydrocarbons in the part per thousand range as opposed to the parts per million measured in open water later) but the buried layers, suggested one team of scientists, were at ten per cent concentration by weight of sand, with perhaps fifteen pounds of oil under each square metre of sand. One kilometre of beach then might contain up to seventy-five tonnes of oil. The team, from the Bedford Institute of Oceanography, estimated, conservatively, that if there were only twelve such beaches in north Brittany 'some 900 tonnes or 4,800 barrels of the *Amoco Cadiz*'s cargo lie within the sandy sediments' ready for potential release.

Even more worrying, and perhaps even more huge in quantity, is the oil which went to the bottom in the sub-tidal areas. Much of the affected coast has a coarse and pebbly bottom continually exposed to strong tidal currents which did not allow the oil to settle. But in many other areas, embayments with fine silt bottoms, weakened currents allowed the oil to be deposited and held.

Two of the largest areas for this were the huge bays of Morlaix and Lannion where, reported Lucien Laubier, summing up the spill in the autumn of 1978, 'The total hydrocarbon content of the subtidal sediment . . . varied from 1–2 per cent, from 24 feet down to 90 feet depth. Assuming that this heavy pollution is general for this area it gives a hypothetical figure of some 40,000 tonnes of oil in the subtidal sediments'.

Underwater TV cameras had revealed that at some places 'there is a continuous covering of oil at the surface of the sediment'. And later in the same summary, Laubier wrote that 'the amount which was dispersed naturally or artificially at sea is difficult to quantify; 40,000 to 50,000 tonnes are thought to have sunk into the subtidal sediments'. In other words, more oil than the total which drifted on to the French coast from the *Torrey Canyon* may have gone to the bottom from the *Amoco Cadiz* spill, much of it in these two bays alone, with future effects that can only be guessed at.

One alarming phenomenon possibly connected with the oil was noted by Chassé's team. During August, six months after the spill, when hydrocarbon levels in the open waters generally had fallen away and most marine life was said to be showing good new recruitment, the scientists suddenly began to discover grey mullet with serious skin ulcerations. The fish were to be found in coastal waters from the Aber Benoit to as far as Lannion. Fishermen told of two other species where from thirty per cent and eighty per cent of their catch were also affected. Other fish, including bass, plaice and conger eel from inshore waters also showed some of the same symptoms: flat fish had damaged fins and some soles, blotches and blisters. Chassé was unable to explain the cause of this phenomenon but he remains convinced it had something to do with oil on the sea bottom.

He is only one of a number of experienced researchers who are worried about the effects of the oil remaining along the north Brittany shores. To take just one of his colleagues, Professor Jean Bergerard, director of the Biological Station at Roscoff, which has been keeping a watch on the local environment since 1372, said he felt that the retention of oil in the sediment was the most serious problem. 'There is a risk of long lasting consequences,' he wrote of a situation in which repopulation of species was already expected to take a long time, 'and therefore one can reasonably ask oneself if (repopulation) will be complete.' And even laymen are struck by the point. The Finisterre Development and Management Committee, while trying to reckon up the impact of the oil on coastal fisheries, sounded its own warning note. While concluding that the immediate effects of the oil did not seem 'as bad as all that', it went on: 'Still, it is too early to say that all the consequences on the living milieu from the catastrophe of the *Amoco Cadiz* are known. Can the reproduction of the species take place as usual in a milieu so disturbed? Only a wide-ranging ecological follow-up will allow us to say with certainty.'

And if there were measurable effects turned up in the ongoing investigations, how to put a price on them? Those responsible for pollution might already find themselves paying up, either through insurance schemes or as a result of national and international legislatory action, for economic damage and for the cost of cleaning up after a spill. But Claude Chassé at least believes it might be also possible to add up the cost of the wildlife destroyed, the fish, the marine animals, the seabirds, the plants, and add it to the bill. Only in that way, he says, will government and the international oil companies come to have any real regard for the impact pollution has on the marine ecosystem.

Chassé suggested that if one took the average mortality of the small herbivores of the winkle family as fifty-four per cent, and that of the limpet family as thirty per cent along 130 square kilometres of affected coast, and assessed the animals as weighing 320 grams per square metre, a total of 16,000 tonnes had been killed. Adding the crustaceans and the populations of the submerged rocks might produce another 25,000 tonnes. A similar operation for the animals living in 420 square kilometres of sediments affected, on a mortality rate of thirty per cent and a weight of between 400 and 600 grams of animals per square metre would add up to 75,000 tonnes (and this would be an under estimation because it did not include the animals washed up from the depths on to the beaches, notably at St Efflam). Together this came to a total kill of 100,000 tonnes of marine creatures.

But the oyster growers had been offered a price for their oysters destroyed of £476 a tonne: 'In order to get an idea of how one costs the replacement of 100,000 tonnes of creatures let us keep that figure. It appears, then that the ecolocial cost of the *marée noire* may be not less than the cost of the clean up by the Army and the others, or over £47 million', says Chassé. And that would be only the loss to 'capital', the stock of such animals along the coast, which might be expected to produce new recruits of between fifty per cent and 200 per cent of existing numbers each year according to species. But because of the oil 1978 was probably already lost, 'so one passes on to £71 million'. Not everything can be costed out in such a way: the birds, though after *Torrey Canyon* some puffins were shipped into Les Sept Iles from the Faroes to build up stock, would be difficult to put a price on.

But, Chassé argues, the structure of the milieu and the relationship between its habitats have been interfered with by the pollution: 'So that it can be defended it is important to proclaim and to show that Nature can and should be valued. In our profit motivated society anything not known to have value is sacrificed, truly disregarded. "Without price", "inestimable" has about as much worth as a grain of dust; and the slide in the direction of scepticism is not only among those who ought to repay.'

11/Wisdom After the Event

LOOKING BACK ON THE LOSS of the *Amoco Cadiz*, it is tempting to speculate on what might have happened if different decisions had been taken. Would the *Pacific* have done more good pulling in the other direction, or at a different angle? Would a second anchor have helped? What if Bardari had used his engines astern earlier in the day, or even gambled on giving them a burst full ahead? Might the tanker after all have been saved if Weinert had connected the second tow to the bows, like the first?

Its a fascinating game to play from the safety of an armchair. But its only positive justification is to help prevent similar casualties.

Before detailing what happened off Ushant on 16 March 1978, we had established that we were dealing with no more than a typical ship on a typical voyage. The question is whether what befell her and her cargo need also be typical. Are such disasters the inevitable, unavoidable price of fuelling our highly geared Western economies with oil, or can something be done to prevent them? Is the black aftermath unavoidably chaotic as well as dirty? And when the cost of all those spoiled holiday beaches, poisoned marine life and choked oyster beds is finally added up, who, ultimately, should pay the bill?

In the boardrooms of the companies with a financial interest in the Liberian tanker, the financial inquest had begun almost before she hit the rocks.

Bugsier had nothing to calculate but the cost of diverting its tugs and replacing lost towing gear – plus the fact that one of its captains was under French arrest. He had, after all, finally succeeded in offering his services on the traditional basis of 'no cure, no pay'.

Shell, which owned the spilled oil, realised with a jolt that its $20 million cargo was not externally insured. But it was Amoco which had to face the most frightening arithmetic.

The ship herself was admittedly insured. So that was all right for everyone but the underwriters, many of them members of Lloyd's in London. But the point here was that initial liability for pollution damage compensation would fall on whoever was established in law as the ship's owner, and that was only all right for Amoco if its liability could be limited at least to the amount covered by its 'protection and indemnity club' of fellow tanker owners, plus whatever the oil companies would chip in from their own pollution fund – that is under the terms of the TOVALOP agreement or the Civil Liability Convention, and through CRISTAL.

The awful scale of the damage done by the *Amoco Cadiz* can be judged from the fact that until she went ashore the oil industry had considered that this voluntary limit of $30 million would comfortably cover any claim likely to be made against it. Yet within six weeks French companies had filed class action in the circuit court of Cook County, Illinois, seeking a total of $250 million in compensation plus $500 millions in punitive damages. Before the year was out, the Standard Oil Company of Indiana and its subsidiaries were facing claims totalling $1.6 billion.

On 13 September 1978, the Republic of France weighed in with a $300 million suit, filed this time through the southern district court of New York. Further locally generated claims followed later that month from the Conseil General des Côtes du Nord and the Union Departmentales des Associations Familiales du Finistère for $366 millions and $200 millions respectively.

Some of these claims were duplicated. Some, no doubt, were inflated by greed; others by the judgement that if liability was eventually to be limited, the highest pro-rata payments would go to those who made the highest claims. But even if only, say, one in ten could finally be substantiated, it would still suggest that the existing compensation limit was hopelessly inadequate. And the first legal issue, on which Standard Oil immediately began its own proceedings, was whether liability should be limited in this particular case, either by international convention or under US law.

The French Government's claim was especially controversial, because France was a signatory of the Civil Liability Convention, which clearly applied to this casualty since the *Amoco Cadiz* went ashore in French territorial waters. Yet the French authorities seemed to be deliberately ignoring their own courts, preferring to test their claim in a country not governed by the Convention, or by its particular concept of limited liability.

The International Convention on Civil Liability for Oil Pollution Damage, to give it its full title, came into force in 1975. Unlike some legal documents, it is written in reasonably plain English:

> *Article II*
> This Convention shall apply exclusively to pollution damage caused on the territory, including the territorial sea, of a Contracting State and to preventive measures taken to prevent or minimize such damage.

Article III
1. Except as provided in paragraphs 2 and 3 of this Article, the owner of a ship at the time of an incident, or where the incident consists of a series of occurrences, at the time of the first occurrence, shall be liable for any pollution damage caused by oil which has escaped or been discharged from the ship as a result of the incident.
4. No claim for compensation for pollution damage shall be made against the owner otherwise than in accordance with this Convention. No claim for pollution damage under this Convention or otherwise may be made against the servants or the agents of the owner.
Article V
1. The owner of a ship shall be entitled to limit his liability under this Convention in respect of any one incident to an aggregate amount of 2,000 francs for each ton of the ship's tonnage. However this aggregate amount shall not in any event exceed 210 million francs.

Standard Oil implicitly acknowledged that the Convention applied to the *Amoco Cadiz* by instructing its insurers to pay $16.7 million into the French courts from which, according to another section of Article V, the money would be 'distributed among the claimants in proportion to the amounts of their established claims.'

The British Government, with vast shipping interests of its own to protect, also took care to abide by the international agreement, and announced that it would be pursuing its own claim for clean-up costs through the French courts. This meant that it could expect to recover only a proportion of the limited sum available for compensation – unless, in the words of Article V's crucial second paragraph, it could be shown that 'the incident occurred as a result of the actual fault or privity of the owner'. In that case, liability would be unlimited.

Theoretically at least, this line of argument was open to the French Government. But both the Paris lawyers and those representing the hoteliers, fishermen and shopkeepers of Brittany evidently decided that their best chance of breaching the liability limit was to ignore their own courts and seek remedy under US law. It looked like the beginning of a long legal battle that would stretch over many years, just as the claims arising from the *Torrey Canyon* had done. Yet ironically, one of the main reasons for setting up voluntary compensation arrangements and international conventions in the wake of that earlier casualty, eleven years previously, had been to cut out this sort of legal wrangling and to ensure that genuine claims were settled promptly, with as little recrimination as possible.

As the writs went in, Standard Oil lawyers began their action to limit their liability under US law and announced that the suits were 'not expected to have an adverse effect on Standard's consolidated financial position.' Then in a legal skirmish started by the French claimants, they seized their chance to protect themselves even further:

On 30 December 1978, attorneys acting for some of the French plaintiffs discovered that the Bugsier tug *Atlantic*, sister vessel of the *Pacific*, had entered the port of Norfolk, Virginia. Within hours they had secured a legal attachment on the £5 million tug preventing her from leaving port.

Their action was intended as a legal weapon to establish jurisdiction over Bugsier in the American courts, for the moment in the US District Court of Eastern Virginia, and it permitted the French claimants to hold the *Atlantic* as part payment of any judgement against Bugsier which might be registered in a US District Court for the company's liability over the part it played in the *Amoco Cadiz* affair. The lawyers for the French told the court that on 16 March the *Pacific* was unseaworthy and that Bugsier had failed to exercise proper care in salvaging the *Amoco Cadiz*. To get the release of their tug, Bugsier gave a letter of undertaking to the court, guaranteeing that they would pay at least £5 million of any proven claims and appear in a District Court in conjunction with the claims.

On 3 January 1979, the *Atlantic* was released. But before she could sail, the tug was re-attached in a new action, this time filed by Amoco Transport and Amoco International Oil Company. Amoco's claim alleged that Bugsier's negligence had caused the *Amoco Cadiz* grounding, that the *Pacific* was unseaworthy at the time of the salvage attempt, and that Bugsier had misrepresented the *Pacific*'s ability to help the tanker.

Most crucial of all, Amoco contended that Bugsier should pay all the damage claims resulting from the spill plus another £25 million to cover the loss of ship and cargo, as well as all expenses incurred by Amoco in responding to the grounding and the oil spill. Two days later, after Bugsier had provided a second letter of undertaking for Amoco, the *Atlantic* was finally allowed to sail.

Amidst all this legal wrangling, the French were absolutely adamant on one central point. In their opinion blame for the entire *Amoco Cadiz* disaster lay with the tanker owners and their ship. 'One cannot repeat sufficiently that the grounding and the pollution were a direct act of the ship *Amoco Cadiz*. That alone was the origin of the disaster,' the subsequent Senate commission reported emphatically. 'That the French authorities had not been able to prevent the grounding and therefore the pollution changes nothing on the exclusive responsibility of the owner and his captain. It was not the French system of surveillance, or of preventing or fighting pollution which generated the damage. It was the grounding, the action of others, which was the author of it.' And the report insisted that to think otherwise would be 'scandalous irresponsibility'.

The bitterness behind the words was felt strongly across France and it was why the French had been quick to bring Bardari before the courts, and, because of the implications of the actions of the *Pacific*, Hartmut Weinert as well.

The French authorities launched two inquests into the disaster. A commission from the Senate started work immediately and reported in one hundred days, and another from the National Assembly, presented a more detailed summary by December. Both inquiry teams asked themselves the same two basic questions: did we have the power to do anything about what happened, and if we did, why were we not able to stop it?

First, they searched the legal texts to see whether French authorities had the legal right to do anything about the ship of a foreign nation with a breakdown at sea and threatening their coastline. And both inquiries were strongly of the opinion that they did. In the international field they found the Brussels Convention of 29 November 1969. Introduced in the wake of the *Torrey Canyon* affair, but not in force until 6 May 1975, it legitimised retroactively, the decision by the Royal Návy to bomb the tanker. But, the French decided it gave clear instructions that nations could take action on the high seas to, in the words of the National Assembly report, 'prevent, attenuate or eliminate grave and imminent dangers presented to their coasts or interests by pollution or the menace of pollution of the sea by hydrocarbons following an accident at sea . . . which may have important damaging consequences.' This convention was important since it changed the whole understanding of the phrase 'the freedom of the seas', gave states unprecedented peacetime powers and did not limit action to ships of signatory countries.

And in their own national law there was the toughly worded law of 7 July 1976, passed in the light of the *Olympic Bravery* and *Böehlen* affairs, Article 16 of which appeared to put anything they wished to do on a legal footing. The article, originally drafted by Guy Guermeur, who now chaired the National Assembly commission, put a legal duty on the owner of a tanker to take all action to put an end to any threat by his ship to the coastline and gave the State powers to order him to take such steps. Where he did not, the State could then act and take measures to recover the costs.

So if they had the legal power to act, whose job was it to intervene? Here they found themselves having to consider who might have had the information on which to act, and who had the means to carry out any intervention which might have taken place. And both inquiries found themselves in a veritable quagmire of conflicting laws, decrees, rules and regulations and habitual practices. They found more than ten government ministries and agencies who had some role, however large or small, in the job of policing, navigation, safety, surveillance and rescue at sea; and another twenty who possibly had something to do with problems of pollution. Each new law only seemed to confuse things further as the different departments fought rearguard actions to protect historic tasks.

On surveillance, a number of organisations from the customs, the gendarmerie, the lighthouse service, the Navy, the port and harbour authorities, to the Merchant Marine, were all supposed to look at some aspect of what was happening off the coast; and the Post Office, through its marine station at Le Conquet, kept a listening watch on some frequencies. But their networks were self-contained, self-centred and in no way complementary in an organised sense. Only the Merchant Marine, with its CROSS search and rescue information centres, and its navigation control centres at Cap Gris Nez and Jobourg, had any kind of specialised personnel or equipment among the civil administrations; and the Navy had its string of signal posts and lookouts for what it insisted were essentially military purposes.

As for intervention, it seemed once again that the Merchant Marine and the Navy legally shared the role of assisting and rescuing those in danger at sea and protecting the coastline. But, taking away its training ships and those of the ISTPM, the Merchant Marine had only nine smallish regional patrol boats and four coastal patrol boats in the whole of France; so for the means to fulfil its duties it fell back on the Navy, which at least had warships, planes and helicopters already carrying out routine missions in the area, and patrolling the Ushant navigation lanes.

If the burden of intervention then fell on the Navy, why did it not intervene? The explanation seemed to be that it was not made aware of the need for action until it was too late. To begin with there was no distress message or direct call for help to the Navy from either the captain of the *Amoco Cadiz* or the tug *Pacific*; and the National Assembly inquiry discovered that things are not done in the same way at sea as they are on land; 'a solitary man, dignified and courageous, the sailor does not call for help until the situation is desperate, until there is peril for human life;' conversely no one ashore, following the same tradition, was ready to make a move until such a call was received. The law of the sea, concluded the report, was confidence in the captain, not defiance. 'After all', said the inquiry, 'every day there are breakdowns, and every day, towing; tanker or trawler, the law is identical;' and the same applied to the reaction of COM at

Brest to Stiff's message to them: no distress appeal, no intervention; 'Such are the traditions of sailors.' That new regulations had been passed only on 9 March, a week before, giving the Navy specific new roles in tackling ships which presented a threat of pollution, seemed not to be a part of the equation. Naval officers gave the impression that protection of the coastline was very much a subsidiary role; indeed a Navy spokesman had been seen on regional television saying so.

Who then, in view of what the inquiries called the 'obstinate silence' of the two ships' captains, should have sounded the alarm and made something happen? The principal witness to the day's events appeared to have been Radio Conquet. Under law, the station's duties were simply those of handling a correspondence service, and, the re-transmitting of urgent messages and distress calls from ships which demanded assistance because human life was in danger. Outside this they had a strict duty to keep secret private communications, even those, like those broadcast between the tanker and the tug on 16 March, to do with aid. But the same decree of 9 March, which had supposedly changed the Navy's role, placed a new responsibility on the radio station to inform the *Préfet Maritime* of anything important happening at sea; yet though the Post Office were aware of the law, the station had had no strict instructions and behaved as before. The signal posts and lookouts of the Navy did not consider themselves, the National Assembly inquiry decided, required to watch over commercial traffic. The drama taking place in front of the watchkeeper at Stiff, even the presence of the *Pacific,* required no detailed surveillance or operation beyond his usual duties. In any case he had not receieved any precise instructions from COM.

The appropriate CROSS centre had specific duties to pass on information concerning the policing of navigation and danger to vessels and life at sea; but it only learned too late to do anything to prevent the grounding.

And it was perfectly clear to both inquiry commissions, that even if one or all of these sources had alerted the Navy (and at least one of the lookouts ought to have queried the presence of such a huge supertanker in difficulties and on an abnormal course outside the navigation lanes) little could have been done. The Navy had no patrol out that day by sea or air; nor did it have call on any specialised personnel who might have been able to help with the breakdown, nor tugs near enough or powerful enough to prevent the grounding.

A patrol ship would not only have told the authorities ashore what had happened but would have enabled the evolution of the towing operation to be followed, the commission concluded, 'The existence aboard the ship of a crew capable of replacing the small crew of the *Amoco Cadiz*, which was already tired, could have reduced the time needed to make fast the second tow. Beside this it seems to us there was the possibility of helicoptering some engineers on to the tanker to try to repair the steering breakdown or at least to define its nature with precision.'

And so it seemed that neither the Navy, nor the Merchant Marine, to whom the premier roles of aid at sea and the prevention of pollution had been entrusted, had the means to do much about the *Amoco Cadiz*. The Navy had only tugs for its own purposes, towing nuclear submarines and ships of up to 30,000 tons, and they were not available. The Merchant Marine had no means except to hire what it wanted from private companies; the only help it could have given was to contact French or foreign tugs in the area and for that it would have needed an up to date list of the tugs available, which it did not have: the presence of the *Pacific* was pure luck.

That a tug of a foreign country was the only means available to help the *Amoco Cadiz* did not say much for the facilities of France, the inquiries decided. But in fact there were only seven tugs of more than 4,000 hp in the whole of France; one of 8,000 hp and two of 16,000 hp. The reports consoled themselves by reporting that other countries in the same situation might not have found themselves much better off. Of tugs over 7,000 hp, there were nine in Norway; five in Sweden; two in Denmark; seven in West Germany; sixteen in Italy; five in Belgium; and seven in the UK. South Africa had fourteen including two of 20,000 hp for just such an emergency.

The National Assembly report was critical of the relationship between the two ships involved, speaking of mounting tensions in the protracted exercise, 'refusals and silences on one side and from the other, evasions;' and it said 'to work properly, aid at sea needs promptness, competence, diligence and harmony. On the contrary the actual atmosphere revealed showed conflicts of interest and of authority and it led to some extremely unfortunate behaviour.' The idea that assistance at sea was simply a private contract was, the report said, 'not only anachronistic but dangerous'. New rules ought to be written so that there would be no need for conflict over terms. Present salvage terms led tugs to go to seek work in an aggressive, piratical way; and for ships' captains to hang on until the last minute before calling them. Indeed, the report went on to examine the question of whether a tug captain might be liable under law for pollution caused by a vessel he was towing. The Brussels Convention certainly looked as if it gave third party victims a right to sue where the aiding vessel had 'aggravated or provoked' the pollution.

But in the final analysis, said the Senate inquiry, whose twenty-one-man commission was chaired by André Colin, president of the Brittany Regional Council, the blame had to rest with the central Government. For while successive responsibilities over navigation control and pollution had been piled on to the various administrations very little had been voted in the way of specific funds to allow them to carry out their jobs. Indeed the total for 1978 was probably only around £2.85 million.

The National Assembly inquiry recorded that between 15 July 1977 when the Ushant navigation lane scheme was introduced until 28 March 1978, the Navy, which was supposed to police the lanes, carried out only 946 hours of temporary watch, 2.5 per cent of the potential hours at sea. The surveillance, which was carried out by sophisticated destroyers and frigates not properly suited for such work, still reported 279 infractions of the scheme, showing the need for patrols.

Similarly, the Merchant Marine had been trying to reduce its reliance on Navy facilities by replacing its fleet of nine patrol boats, which were unsuitable for bad weather, with a new type, the *Garance* class of 16,000 hp and capable of twenty-three knots. But of seven planned, only three were in service. In addition the Government had only just voted funds for the first stage of the badly needed Ushant control centre to match those for Jobourg and Cap Gris Nez.

Nor was pollution fighting cheap. After the sinking of the *Böehlen* in 1976, the Navy had had to keep a watch on the resulting pollution and treat it for ten months. Together with the pumping operation by private contractors, the bill had come to £18.9 million. Of this, £18.2 million was eventually recovered but that still left the Navy with some £714,000 to find from its own budget.

The Senate inquiry concluded that a change of Government attitude was the first priority, from thinking of pollution as some kind of unpredictable fatality to realising it was a continual danger against which proper preparations had to be made.

And, to the consternation of other nations, the French Government did take swift unilateral action to ensure that nothing like the *Amoco Cadiz* happened again: imposing regulations, with effect inside and outside her territorial waters, unprecendented in peacetime. The Navy was ordered to go over to a twenty-four hour watch by sea and air. On 24 March came Prime Minister Barre's announcement of new controls of navigation for tankers. Initially, tankers were instructed to keep within two miles of the outside edge of the separation lanes; and at the same time all but local traffic was banned from the channel between the island of Ushant and the mainland. Ships were also ordered to keep in touch by radio, and radar sets to watch over the ocean were hastily installed as a temporary control centre in the lighthouse at Créac'h. Then on 22 May new navigation instructions were issued by the *Préfets Maritime* covering the whole of the French coastline from the English Channel to the Spanish border. They laid down rules for tankers entering and leaving harbour: masters were required to report their movements in territorial waters and send reports of their capacity to manoeuvre and navigate six hours before entering territorial waters or leaving port.

At public expense, the French put one of the two biggest ocean-going tugs in France into Brest, the *Abeille Normandie* of 16,000 hp, capable of sixteen knots and with a pulling power of 120 tons: nearly twice the power of the *Pacific*. Three new radio beacons and a radar buoy were installed in the Ushant area (and another, marking the entrance to the northbound lane, standing thirty-nine feet out of the water, was to be laid in spring 1980). The Ushant area was already covered by Decca Navigator electronic navigation stations, allowing ships which had the facility to track their position to within one-tenth of a mile over ninety-five per cent of the area, and to within one mile elsewhere; and in May it was supplemented by the OMEGA satellite navigation service to give even more precise coverage.

A new corps of survey officers, to check on the conditions and qualifications of tankers and crews entering and leaving French ports, was instituted. On 2 August, a special new Interministerial Mission for the Sea, headed by a Minister for the Sea, Achilles Fould, was created to co-ordinate the pollution work being undertaken by the different ministries.

By October 1978 a revised version of POLMAR had been drawn up, leaning heavily on the experiences of the *Amoco Cadiz*. It placed responsibility for all action at sea firmly in the hands of the *Préfet Maritime* and at the same time handed back control of the land operation to the Ministry of the Interior and its Civil Defence organisation, a clear reflection of the contribution of Becam and Gérondeau. The new scheme revised the roles of the ministries and laid down clearly the parts to be played by the two sides of the plan. Co-ordination with local leaders was institutionalised. CICOPH was to be replaced by a new organisation, CEDRE (*Centre d'Etudes de Documentation de Recherche et d'Experimentation*) which would be based at the Brittany Oceanographic Centre, taking a staff of twenty from other organisations, to research and advise on pollution fighting. New plans for local intervention included lists of sensitive zones to protect; materials held; precise plans for barrages; sites for intermediate and final storage of oil cleaned off a shore; treatment centres for waste; a plan for the transport of commercial shellfish; accommodation for POLMAR personnel and volunteers; a list of trained staff; and new arrangements for financial powers and intervention funds.

Air patrols were told to augment the Navy patrol ships and in January 1979 the laws covering infractions of the navigation regulations and polluting the seas were tightened to provide, in the case of navigation rules, for fines on the master from £5,900 to £119,000, and for negligent pollution from £59,000 to £595,000.

Formidable as this list was, it did not still all the critics, particularly in Brittany. The stationing of the *Abeille Normandie* at Brest came under attack, with the suggestion that a contract with a private company, in which government and owner would share salvage prizes, was simply perpetuating the piratical image of the tug business: what was needed, it was said, was a team of state-owned intervention tugs stationed at different points along the French coast. The *Abeille Normandie* was costing £2.38 million a year to charter (plus £1,500 a day when working and a salvage share) when a new tug might cost only £7.1 million.

Late in 1978 the Navy carried out an exercise in which the 259,000-ton tanker *Esso Provence* was supposed to have broken down with engine trouble off Ushant; the *Abeille Normandie* and a Navy frigate set off for her, while a Super Frelon helicopter flew in the Navy's newly-trained intervention team on to her deck. But while the authorities seemed happy with their practice mission, local seamen pointed out derisorily that the exercise had been carried out with an empty ship, in good weather and an almost flat calm. And they queried the wisdom of stationing a sixteen knot tug at Brest when the average winds off the island of Ushant are notoriously high. The *Abeille Normandie*, even if alerted, might arrive too late to help a ship in danger. So the locals suggested building a dock for her at Stiff, on Ushant itself.

There was another argument about whether the new penalties for breaking the navigation rules were swingeing enough. The National Assembly report itself suggested seizing the infringing ship, with the owner having to prove his vessel was not at fault before he could have it back. 'The polluters of the sea are the pirates of modern time. The sanctions against them must be drastic,' it said militantly. But the Government could not be persuaded to be firmer; the official attitude was that in addition to the punishments laid down in the laws, any ships arrested would also lose at least one or two days running and operating costs while in port and also run up port and pilotage charges, making a substantial surcharge that would make masters think twice.

Already the new surveillance had had its effects. In the four months from

May to August of 1978, the Navy had put in 3,350 hours of sea patrol and thirteen by air; and backed up by the new radars at Créac'h, and an improved communications system between COM and the Stiff and Créac'h lookouts, had brought down the number of infringements from twenty-two a day in April to only nine a day by September. Arresting ships and taking them into port as examplary cases had also cut down on repeat infringements.

Where radar surveillance was concerned, the French ran into troubles of their own making. From 1 January 1979, with the reluctant agreement of IMCO, the shipping lanes were changed to keep loaded tankers well clear of Ushant. The radar installed at Créac'h had a range of only twenty-six miles while the inbound tankers were being asked to keep at least twenty-seven miles out. Most tanker traffic would therefore be on the limits of the coastal station's electronic vision. A new £4.1 million sophisticated tracking radar due to be installed in a new traffic control centre at Stiff early in 1980 will have a range not much better than thirty miles. The National Assembly report calculated that to get most of the tanker traffic in satisfactorily, the radar would have to be on a tower 465 feet high. To get another eight miles range it would have to be 786 feet high; and to cover fifty miles it would have to be an enormous 1,518 feet high. It was clear that to get practical coverage a radar platform might be needed at sea; but this would be costly and would provide yet another hazard to navigation in an already treacherous area.

In any case prevention was only half of the equation and central to all the discussions about what had been done was whether the Government had solved the problem of putting together an effective response to an oil spill if another tanker went aground on French shores. If the system for intervening at sea was complex and compartmentalised, said the Senate inquiry report, then POLMAR was worse, a two-headed machine that had taken six days to lumber into place. And although a co-ordinator had been quickly found, the organisation simply did not, to begin with, have sufficient means at its disposal. To take only one point: if the honeywagons had not been available POLMAR would have had nothing, within a reasonable time, to bring up to pump the oil. The Senate warned that 'tomorrow, in the same circumstances as those of 16 March, the same causes would have the same effects; and to hide it would be to tell lies to the nation.'

Nor was the National Assembly report happy about the revised POLMAR plan: there was still room for confusion of roles between the different administrations it thought, and, astonishingly, still no plan for vital advance command posts such as those which had been set up at Ploudalmézeau and Lannion.

But the central worry was still the complexity of organising co-ordination between government departments in an emergency; and there still appeared to be confusion over command. In particular there was a large question mark over the usefulness of Achilles Fould's new Mission. With a minuscule budget for its job of only £440,000, the Mission appeared to have no authority or power. Its main strength was limited, 'stingily' thought the National Assembly inquiry, to only twelve officials who were not permanent staff but only on loan from their departments. And, most important, it seemed to be under orders to hand over control of co-ordination in an emergency to the Ministry of the Interior.

Clearly something much more far-reaching was needed. When the two commissions came to decide what should be done, their main conclusion was the same: there should be one central authoritative organisation, to patrol the coastal waters and with powers to intervene at sea at the earliest possible moment to prevent danger from any ship developing; and in the event that that could not be done, to take firm command of a pollution fighting exercise off shore. But the commissions differed dramatically in their choice of organisation. The Senate inquiry, reacting within weeks of the grounding of the *Amoco Cadiz* thought the Navy should be given the role; backed up with funds, men and equipment, it would be a simple extension of the Navy's job of protecting the integrity of the coast from invasion. But the National Assembly commission, after six months of consideration, came down in favour of an entirely new national Coastguard Service, based on the example of the United States, independent of, but funded through the Ministry of Defence.

The National Assembly inquiry was shocked by the gaps in the arrangements for maritime surveillance which had been shown up by the events of 16 March: 'That a ship with broken steering was able, in the indifference and inertia of organisations, persons and authorities normally charged with surveillance, to drift all day in a zone where those administrations were strongly implemented and finally ground on rocks known universally to be very dangerous, gives the measure of the delay to the reaction,' its report said.

And the reason for the inertia was plain: the only authority alerted to what was going on, the Navy, did not take action because it was insufficiently aware of the importance of its mission to protect the coast against pollution, which it did not see as its essential task. It reflected a traditional attitude at sea: that the captain is master and that nothing is done in the absence of a signal of distress. But the authorities' failure also reflected their complex administrative structure, (unlike the US coastguard), and the mediocre nature of French resources. The Navy did not have the means to act or the cash to buy them, and the Merchant Marine had to rely on the Navy. Despite having taken legislative action, the Government had not come up with supportive funds and kept quoting the existence of the TOVALOP shipping agreement as reason for its parsimony. Something ought to have been done after the lessons of the *Olympic Bravery*, the *Böehlen* and the *Urquiola*.

It was because of this attitude by the Navy that pollution and navigation policing was a subsidiary role that the inquiry had come down in favour of a Coastguard, said the National Assembly commission; particularly after watching a Navy spokesman talk on the subject on regional television. The commission said it neither wanted to reduce the defence of the nation nor have a precarious defence of the coast. The danger was that in some emergency there might be an effective 'brutal dismantling of the navigation zones because the Navy ships protecting the coast had been called to other missions.' Its recommendation of a Coastguard Service was a unanimous one.

Working up to the establishment of such a service, the inquiry thought, there should be a global programme for improving surveillance along the coast, with the strengthening of facilities in the CROSS centres. There should be a twenty-four-hour watch from the semaphores which would require another 100–150 men in the Brest naval region. It thought the Coastguard should get air patrol planes such as the Nord 262 and Mystère 20G and ships of 1,500 to 2,000 tons doing eighteen to nineteen knots and strongly built and able to carry a Dauphin helicopter; with fire-fighting, anti-pollution equipment and frogman facilities. There should be smaller boats of 250–350 tons for shallow water work. To back up the service there would have to be revised laws on radio communication to CROSS stations and Navy or Coastguard operations rooms to be told of messages other than distress calls, which would probably mean another thirty staff being put into marine radio stations to cover this duty.

The Senate commission had also looked at the Coastguard option closely, before deciding against it. It noted that the US Coastguard had 40,000 men, 5,000 civilian technicians, 76 patrol ships, 7 tugs, 53 aircraft and 72 helicopters. And it noted especially that the service had three trained teams, in the Atlantic, the Gulf of Mexico and the Pacific, with air transportable equipment, ready to be flown in to the site of an oil spill; that it had an equipment testing centre, and a laboratory to analyse oil samples. But in the end, the inquiry report said, the Navy should do the job since it was an extension of its defence role. 'Military ships are not alone in being dangerous; certain commercial ships today consistute a considerable menace for the national territory and for the civil populations of certain French regions.' The idea of a Coastguard protecting 3,273 miles of French coastline, and a zone stretching 200 miles out to sea, in an integrated permanent structure was an attractive one, but the cost would be prohibitive when the Navy already possessed some of the facilities it would need.

Unlike the National Assembly report, that of the Senate gave a detailed breakdown of the kind of equipment it thought the Navy would need to adopt its new role, together with estimated costings. It foresaw the need to set up a research team like CEDRE and for an intervention unit which could be flown out to the deck of a stricken tanker. But it wanted much more: there should be a new fleet of patrol boats capable of twenty-five to twenty-eight knots, backed up by helicopters and planes; a special tanker equipped to lighten the load of a grounded tanker should be found, ready to intervene within twenty-four hours rather than the days it took to get together the equipment for the abortive attempt on the *Amoco Cadiz*. The inquiry called for more skimmer and mechanical clean-up vessels, suggesting a fleet of at least twenty-five small boats capable of collecting 200 cubic metres of oil an hour. The inquiry also looked closely at a much larger system being developed by an organisation called Chantiers de France-Dunkerque. Outside Britain, the system developed by the company, which is part of the large Empain-Schneider group, had attracted a great deal of serious attention and in Brittany was even being supported by sceptical biologists like Claude Chassé. It centred on a large factory ship, capable of twenty knots, which would carry great lengths of floating barrage, and a catamaran on which a huge pump was mounted. In an oil spill emergency the factory ship would stand off the wrecked tanker, and, using locally hired tugs, put out its barrages to corral the spilling oil. The catamaran would sit at the end of a funnel formed by the barrages and pump the melange directly into the factory ship. There separation machinery capable of dealing with 6,000 cubic metres an hour could extract the oil and pump it along a floating pipeline to a tanker standing nearby. The company claimed the unit would work in winds up to Force 5. In the case of the *Amoco Cadiz* spill, it claimed, a unit based at Brest could have been at work within three hours (and there was a useful lull in the heavy weather in the first days immediately following the grounding which would have enabled it to work hard) and a second, based at Cherbourg would have been on the scene in seven hours. Each unit might have treated 20,000 tonnes of oil in ten hours. Two units, suggested the Senate inquiry, might protect the whole of the French coastline, even though each would cost £23.8 million.

When the Senate commission finally added up the cost of all its suggestions in men, boats, tugs, helicoptors, ships and materials, the total bill came to £595 million at 1978 prices, a staggering cost when compared with the £2.85 million the Government had been planning to spend before the *Amoco Cadiz*.

As to the fight on land, both inquiry commissions were agreed that it needed a single organisation not only to co-ordinate, but to command directly. And they seemed agreed that the task should be handled by the Ministry of the Interior acting through the Civil Defence organisation. The Senate commission accepted that this would mean the abandoning by some ministries of traditional roles and a complete 'change in mentality', as well as being costly, but it believed it was the only way forward. 'Since 1967 the year of the *Torrey Canyon*, all has happened as if one was hoping that an accident of the same nature would not be reproduced,' said the report. And that was the first mentality that had to be changed. Special funds would have to be voted to the Civil Defence, and the National Assembly inquiry thought some of them should be spent on a new 600-man Pollution Commando, based on the existing UISC7 Civil Defence unit which had been in action in Brittany.

Both commissions laid heavy emphasis on a point long argued in Brittany itself: that the solution to pollution problems did not lie with France alone. Around 500 million tonnes of oil destined for various parts of Europe and Scandinavia came past the shores of France each year and it was only right that other countries, particularly those nine in the European Community, should bear part of the load, both financially and in the researching and organising of intervention techniques. Nor could the oil companies escape their responsibility. One way of raising money was a tax on each tonne of oil imported. This had enabled South Africa to mount an effective response including her two huge ocean-going tugs always on stand-by. If a tax of 59p per tonne were placed on French oil imports, that would have raised £70.2 million on 1977 imports, the Senate inquiry calculated. If this were extended across the Common Market, which imported 497 million tonnes of oil in 1977, it would have raised £297 million – well on the way to the inquiry's estimate of the costs of an ideal French operation of £595 million. And any oil company willing to lose a cargo worth £11 million, which was not insured, could certainly bear their share of the cost of the programme, said the Senate commission report grimly. Whatever measures were taken they were needed 'as quickly as possible, for tomorrow it may be too late.'

For as the French inquiries emphasised, the *Amoco Cadiz* was never likely to be the last tanker to come to grief off Brittany. As if to reinforce the point, only thirteen months after the disaster, on the night of 27 April 1979, the Liberian tanker *Gino* collided with the Norwegian chemical tanker *Team Castor* 25 miles West of Ushant and sank with more than 30,000 tons of carbon black in her tanks. Fortunately, the water that distance offshore was 300 feet deep, the ship was still more or less in one piece when she went down and carbon black – a refined petroleum product with the consistency of toffee at low temperatures – is heavier than water. But eventually it will begin to ooze from the rusting tanks, and scientists can only speculate as to the damage it will one day do.

The *Gino*, like the *Amoco Cadiz*, could have sunk off the shores of Britain, of Belgium, Holland, Germany, Denmark or any of the coastal countries receiving a share of the nearly 500,000 tonnes of oil a year which round Ushant in tankers using the 'Route de Cap'. And a second disaster on a supertanker scale would still catch them relatively unprepared, in spite of the reviews of contingency plans which have taken place with some urgency since the *Amoco Cadiz*. For the truth is that no European nations (indeed, to widen the point, precious few other nations in the world) have yet put into the anti-pollution fight the kind of resources needed to give something near real protection. And outside Europe, even in those countries which seem to give the problem a higher priority, such as the United States and South Africa, no one is bold enough to say there have not been lessons to be learned.

In June 1979, in a repeat of the Santa Barbara incident which alerted the American public to the problem of oil pollution ten years earlier – but this time on a much larger scale – the *Ixtoc I* well blew out sixty miles off the Mexican coast, releasing thousands of tons of oil a day. It was the beginning of a long struggle to protect the Gulf's fishing grounds and beaches, about which the most consolatory thing one can say is that there was plenty of warning of the oil's arrival.

Within the next few weeks Caribbean beaches were threatened by the aftermath of a terrible collision between two 200,000 tonners, the *Atlantic Empress* and the *Aegean Captain*, in which 28 seamen died, while on the other side of the Atlantic anxious hoteliers in the English seaside resort of Sandown, Isle of Wight, protested at the appearance of the German tanker *Tarpenbek* anchored just off their pier – upside down.

This last incident ended happily thanks to the skill of the Dutch salvage team which handled it, but it left no one reassured; it was too reminiscent of the upturned hulk of the *Eleni* V, which drifted up and down the North Sea coast for almost a month before the authorities admitted there was nothing the could do with either the ship or her cargo of heavy fuel oil except to blow them up. It was a time for official admissions, including the fact that the national pollution contingency plan for the Channel could initially cope with no more than 6,000 tons of oil a day. The House of Commons Select Committee on Science and Technology pointed out that this bore 'no relation to the likely size of any pollution incident, and leaves the South coast virtually unprotected'.

The Government answer to that was the setting up in November 1978 of its new Marine Pollution Control Unit, to lead the fight against oil spills at sea, with a central commander in the shape of a Rear-Admiral; and the launching of a review of its contingency plans for a clean-up operation. But many local council members and local government officers are still far from convinced that the review is fundamental enough. And the budget after the overall changes already made is still, for example, only about the same as the cost of one of the oil-clearing factory ships designed by Chantiers de France-Dunkerque, and pressed on the French Government by its Senate inquiry into the *Amoco Cadiz*.

And there is little need to single out one country for such criticism. Britain's spending on such contingency measures, like that of many nations, is still based on the gambler's odds that large scale disasters come rarely. But the odds can change. One French authority wrote only a few years ago that the chances of a fully-laden supertanker running on to the Brittany rocks were about one vessel in one thousand years; but more recently it has been suggested inside Brittany itself that the true calculation should be one every decade, which ties in with the British study quoted in Chapter Four.

Yet the French estimate of £595 million for an anti-pollution system not often used is still far from cheap. And the question is whether it is necessary for each country in Europe to build the kind of massive armoury foreseen by the French studies; or whether, in the spirit of the economic links of the European Community and the defence ties of NATO, the resources needed to face pollution might not be shared. For oil is only one, and not even the most dangerous, of the polluting products carried at sea.

One of the sadnesses of the *Amoco Cadiz* operation was the limited international support offered to and sought by the French in their troubles. The idea of joint action has long had strong support in Brittany, where the local people do not see why they should always pay the price for disasters involving oil not even intended for French refineries. And since the *Amoco Cadiz* it has gained ground in national government and EEC circles. One of the first improvements the British made to their system after the events was to lay new emergency telephone and telex links across the Channel. Would it have helped if there was a daily up-dated list of ocean-going tugs on call in the Channel? If both the British and French had special intervention vessels on patrol? If troops with equipment could have been flown in from France and elsewhere as part of a pre-determined plan in the first days? If French and, say, Dutch oil-cleaning factory ships had been able to sail for Brittany in the first few hours? If internationally-held stocks of pumps, pipeline, skimmers, barrages and road tankers, and the expertise in how to use them, could have been raided as the oil came ashore? If more boats with more spraying equipment for dispersants, or even spraying aircraft, had been available? It is because there are no positive answers to such questions that governments usually act only after the disastrous event.

The official Liberian verdict on the loss of the *Amoco Cadiz* came almost a year later in the interim report of the marine board of investigation chaired by Sir Gordon Willmer. This is the inquiry that took its evidence in public at the Royal Institute of Naval Architects in London and it put the main burden of criticism on the design of the steering gear and on Captain Bardari for his 'inexcusable delay' in calling for assistance.

The board accepted that the overriding cause of the casualty was the steering failure, and that when the damage proved irreparable, the tanker's eventual loss may well have been inevitable. Even so, it argued that the captain should have called for tugs as soon as he had made his own inspection of the steering compartment at 10.05, that is about an hour earlier than he actually did make his first call for assistance:

> With the vessel drifting rapidly towards the shore the situation was one of extreme peril, and every minute of time saved was vital. Had the *Pacific* arrived in the area an hour earlier than she did she would have had additional time and sea room to manoeuvre for taking the vessel in tow.
>
> Moreover, it is probable that the tug *Simson*, which in fact arrived only after the *Amoco Cadiz* had already grounded, if she had been alerted earlier, could have arrived when the vessel was some distance from the shore, in which case it is possible that with the combined efforts of the two tugs the vessel might have been prevented from stranding.

Although the board commended the Italian Master for his courage in staying aboard his sinking ship, it recommended suspension of his certificate pending fuller investigation of the part played by his owners' representatives in Chicago. The interim report commented that Bardari showed 'a strange and lamentable reluctance to assume responsibility, as Master of the vessel, for deciding what steps should be taken.'

Nothing in a tanker captain's routine training and experience prepares him for this kind of emergency – though perhaps it should, just as airline pilots practise dealing with engine failures, fires and so on – and with the radio available it must be tempting to ask the advice of head office, indeed sensible if there is really time to do so. Captain Bardari maintained in evidence to the Liberian inquiry that he always had full authority to negotiate with the German tug but he certainly spent a lot of time on the telephone.

At least two big tanker operators thought it necessary to clarify and re-emphasise their masters' responsibilities in the light of what happened that day. In a circular letter dated 5 April 1978, the Shell Tankers management wrote:

> It must remain within the sole judgment of the Master as to the degree of danger to which his vessel is exposed as the result of any breakdown; that judgment must take into account a number of considerations, including proximity to danger, weather and other conditions which may affect the situation, but ultimately the objective must be to preserve life and property, and to make the optimum use of any assistance which may be available to secure that end.
>
> In these circumstances, where after appraisal, an element of real risk is judged to be present, you should not hesitate to accept assistance offered under Lloyds Open Form of Salvage Agreement. Acceptance of these terms does not connote substantial or inflated settlements, but in fact normally results in an objective analysis of the circumstances, leading to an acceptable settlement.

Not that the Liberian board took much account of the haggling over salvage terms that went on between Weinert and Bardari. Its report found their argument unnecessary, unseemly, but largely irrelevant since it accepted the German's evidence that the *Pacific* was already pulling with as much power as she could safely apply.

As for the steering failure itself, the interim report concluded that the gear's design pressure of 140 kg/cm was simply inadequate to take the enormous pressures forced back through it by waves striking the forty ton rudder. Eventually these must have hammered their way through the interior of the hydraulic distribution block to blow out the relief valve pipe (the second pipe failure, and the one which made it impossible to use the system) although it was impossible to verify this assumption without recovering the gear from the sunken tanker:

> But it must be said that the weather, though violent, was in no way exceptional, and no more than might reasonably be expected in this locality in March. It appears to the Board that the effect of such by no means abnormal wave forces

> and the pressures resulting therefrom cannot have been sufficiently taken into account in the design and manufacture of the steering gear. Moreover, it is to be observed that, although the steering gear appeared to comply with the relevant regulations in force at the time of the vessel's construction under the Safety of Life at Sea Convention of 1960, the events which actually happened proved all too clearly that she did not in fact have an effective auxiliary steering gear.

No blame was attached to the ship's engineers, who did 'all that could reasonably be expected of them', or to her builders and owners in isolation, since she had been accepted for classification by the American Bureau of Shipping. Since the casualty, however, all sides of the industry have been re-examining steering gears and other systems that may have outgrown their strength when applied to VLCCs.

As we have seen, the first impulse of the French maritime authorities was to blame the IMCO routeing system. The *Amoco Cadiz* was bound for the English side of the Channel to lighten her cargo in Lyme Bay, so why not go straight across? It was bad enough having a rule that Channel shipping should 'drive on the right' – which meant that all the loaded tankers came up the French side – without drawing routes on the chart that invited navigators to hug the coastline at a dangerous place like Ushant.

It was a fair point. Captain Bardari himself said that as soon as he shaped a course from the Canary Islands he knew he would be passing no more than 7½ miles from Ushant, and when asked why, he replied that he always intended to be 'in the traffic course of IMCO' – he was simply 'making the normal course'. Indeed unless he stayed clear of the Brittany peninsular altogether he was bound by the new 1977 collision regulations to follow the inshore traffic lane.

But of course Ushant used to attract shipping long before there were such things as IMCO traffic separation schemes. The island is an obvious landfall after crossing the Bay of Biscay, a corner ships must turn before heading up Channel. The one-way routes were merely put there to reduce the risk of collision by dividing the existing streams of traffic. And even though the *Amoco Cadiz* was bound for the other side of the Channel, Ushant did prevent her taking an absolutely direct course to Lyme Bay. Captain Bardari had intended to come slightly right, from a course of 032 degrees to 037 degrees, as he left the island astern.

The real issue, therefore, was whether the route for laden tankers could be moved further offshore, so that if they did collide or break down, they would have a reasonable amount of sea room. After an initial political panic, when the French declared that ships carrying oil must simply keep away from the Breton coast, they came up with an ingenious proposal to create a third traffic lane, for inbound tankers only, outside the existing two. The outbound lane would also be sub-divided, so as to keep tankers on the offshore side of it.

The effect of the change was dramatic. It moved the centre of the lane used by most loaded tankers (only a few have reason to carry oil down Channel, although this number will increase with the development of the North Sea oilfields) from seven to more than thirty miles offshore. IMCO accepted the idea, but not without some anxious debate, because apart from making position-fixing more difficult, for this important class of ship it meant a departure from the basic principle of keeping to the right, which is reflected in the steering rule that vessels meeting head on should turn away to starboard. It was the sort of expedient working compromise that the British might have been expected to propose, rather than the logical French, and there was an uneasy suspicion that IMCO was being stampeded into an unseamanlike solution by political pressure – however understandable that pressure might be.

Trinity House, the British pilotage authority, openly condemned the French scheme as dangerous, on the grounds that after leaving Ushant, supertankers would have to cross the oncoming traffic so as to join the Casquets separation scheme off the Cherbourg peninsular on the normal right hand side. The Casquets scheme was also being moved further out, with the inbound lane sub-divided as between dry cargo vessels and tankers to keep the latter on the offshore side. But there was no room there to create a third tankers-only lane and even if there had been it would simply have moved the crossing problem further up Channel.

Finding its protests about the changes at Ushant ignored, Trinity House produced a much more drastic plan to replace all the piecemeal separation schemes by a single pair of through routes running from one end of the Channel to the other, right down the middle. Here the appropriate analogy was not with the airways, but with motorways ashore. There would be only three or four 'intersections', marked by buoys or lightships, where ships would normally be expected to join the one-way lanes or cross them.

It was an attractively simple concept. It certainly met the French authorities' desire to keep tankers away from their shores. But it was bound to be deeply controversial because it would encroach on the coastal seaman's freedom of navigation as well as the deep sea sailor's. If the existing collision regulations were applied to the broad lanes sketched out in the original Trinity House proposal they would also interfere badly with the Channel fishermen – who find the whole business of routeing schemes a nuisance anyway.

IMCO regulations state that: 'A vessel engaged in fishing shall not impede the passage of any vessel following a traffic lane.' This restriction already causes trouble in the short lengths of one-way routeing that exist, let alone in a continuous scheme drawn the whole length of the Channel, right through the offshore mackeral grounds. Such a scheme would also be awkward for yachtsmen, many of whom find navigating the Channel in their flimsy craft difficult enough as it is, without trying to obey traffic rules written primarily for large power-driven ships.

One of the first casualties of the new era of navigational control was the old principle that steam gives way to sail. In the open sea a sailing vessel still does have the theoretical right of way, though no yachtsman who valued his life would try to assert it. Modern ships move too fast. They cannot be relied upon to keep a visual lookout and by the time a small yacht shows up on a radar screen it may be right under the oncoming vessel's bows.

This much has long been accepted, because it was dictated by self preservation. Yachts have always tended to avoid known shipping lanes for the same reason. But when the new collision regulations came into force the sailing fraternity slowly woke up to the fact that they too were covered by the traffic rules, however small their craft. The giant tanker might be restricted, but she had at least acquired commensurate privileges when manoeuvring her great bulk through shallow channels or following a mandatory traffic lane.

All vessels are bound by Rule 10 to cross separation schemes 'as nearly as practicable at right angles to the general direction of traffic flow'. The last paragraph adds that: 'A vessel of less than twenty metres in length or a sailing vessel shall not impede the safe passage of a power-driven vessel following a traffic lane.'

Yachtsmen had read these words without knowing how the authorities would interpret phrases like 'as nearly as practicable,' or 'impeding the safe passage of a power-driven vessel,' where they were concerned. They began to find out in 1978, when the British Department of Trade, acting on Coastguard reports, prosecuted three yachtsmen – two of them for not crossing the Dover Strait at right angles and a third for sailing gently up the wrong lane.

At first sight it looked like a classic piece of armchair sailor's bureaucracy. After all a twenty-five foot boat was not going to make much of a dent in the bows of a 1,000 foot supertanker. But when the yachting press asked for clarification of the official policy, they were met with careful argument and a good deal of somewhat unexpected sympathy. The Coastguard pointed out that quite a high proportion of the 'rogues' still showing up on the radar screens were yachts. During one week in the summer of 1978, nineteen of the seventy-four rogues that were positively identified – as opposed to being plotted simply as anonymous blips – turned out to be pleasure craft. The problem as the Department of Trade saw it was not so much the possibility of a direct collision as that of a ship risking collision with another big vessel by giving way to a small yacht floundering in its path. Those sifting the Coastguard reports were well aware of a sailing boat's limitations, it seemed, particularly in bad weather. They were happy to see a yacht 'tacking' to and fro in the middle of a shipping lane, provided she was going in the right general direction. But they would not accept that she was entitled to ignore the traffic rules altogether just because she was small and slow.

There is clearly room for a practical compromise here. But the fact remains that the weekend sailor's freedom to wander wherever the breeze takes him is being progressively curtailed, just like the professional mariner's freedom to plot the shortest course between one port and another.

Eventually, something like the maritime 'motorway' envisaged by Trinity House may well be laid down right round the British Isles, and perhaps along other busy sea routes such as the eastern seaboard of North America. How quickly it appears and how far it extends will depend on whether seafarers can evolve a workable set of rules for it that will reconcile all the different interests that are at stake – a slow process, as IMCO repeatedly demonstrates. A first step might be to move the Casquets separation scheme into mid-Channel and link it to the Dover Strait scheme by a recommended route – not a compulsory one, just an obvious route that navigators will want to take because it is so well marked by buoys or lightships, and perhaps indicated on the chart by a broken line instead of a continuous one. In other words the motorway analogy could be extended to include the various lines that are used on roads to indicate differing degrees of lane discipline – provided no one ever forgets that you cannot paint lines on the sea.

Meanwhile the shore-based surveillance and control of Channel shipping will continue to develop in the form of the British Coastguard's Channel Navigation Information Service (CNIS) and its French equivalent, the Service d'Information et de Surveillance de la Navigation en Manche (SINM). There are plans to extend the radar coverage at both ends of the Channel, for example, and now the CNIS have their purpose-built control centre at Langdon Bay, packed with electronic equipment that can work miracles of instant analysis, they are bound to think of new ways to use the information coming in. And here, of course, the comparison must be with an air traffic control centre.

Radar transponders are an obvious possibility. They are fitted to airliners so that the ground controller can confirm which of the blips on his screen represents the aircraft in which he is interested. The pilot identifies it by switching on his transponder, or 'squawking', when requested by the controller, and clearly a ship could do the same.

Whether or not transponders are used, there can be no positive control of shipping traffic unless the individual vessels identify themselves. The reporting-in scheme which began experimentally in 1979 is therefore crucial to the system's long term development as well as being part of the immediate political reaction to the *Amoco Cadiz*.

The trial scheme applied to four categories of ship:

(a) all loaded oil tankers and loaded gas and chemical carriers of 1,600 gross registered tons and over

(b) any vessel 'not under command' or at anchor in a scheme or associated inshore zone

(c) any vessel 'restricted in her ability to manoeuvre', vessels engaged in towing, deep draught vessels, or vessels with a defect in their propulsion or steering

(d) any vessel with a defect in those navigational aids which could adversely affect her navigation under prevailing conditions.

In each case, except for the anchored vessels, they were asked to make radio contact with the appropriate shore station ten miles before entering the traffic schemes at Ushant, Casquets and Dover, giving their name, position, course, speed and intentions. The actual wording of the British Notice to Mariners 'invited' masters to report in if they wished to do so, but the French authorities adopted a more robust attitude, as they have done to the whole business of policing the traffic schemes. Within French territorial waters they have imposed their own regulations on top of the IMCO rules. Nor have they been content simply to report traffic violators to their respective flag states. A number of foreign vessels have been ordered into Brest roads by the French Navy and not allowed to leave until a fine was paid.

In practise, ships taking part in the experimental reporting-in scheme also declared their cargo, although this was not actually required by the rules. The reason for such a curious omission – a combination of commercial security and Soviet pressure inside IMCO, no doubt prompted by the other kind of security –illustrates the way in which limits are set to this kind of control even in such dangerously congested waters as these.

The only test of shore surveillance and routeing control which ultimately matters, is whether they have reduced the risk of collisions and other casualties in the Channel, particularly those causing heavy pollution. And the initial evidence is encouraging. The number of collisions in the Dover Strait fell from a peak of thirteen in 1971 to only three in 1977 and four in 1978.

THE DOVER STRAIT

Year	*Number of Collisions*
1971	13
1972	6
1973	3
1974	7
1975	5
1976	5
1977	3
1978	4

In any case the large reduction in the number of ships moving against the main traffic flow must have had some effect on the number at risk of collision, even though the separation schemes down Channel cannot match the Dover Strait's well disciplined record. An Anglo-French survey in June 1977 – that is before the French stepped up their police work in the wake of the *Amoco Cadiz* – showed two per cent of 'rogues' amongst the through traffic in the Strait, twenty-seven per cent off the Casquets, seventeen per cent at Ushant and no less than forty-two per cent off the Lizard. These figures probably demonstrate the deterrent value of the close radar surveillance from St Margaret's Bay and Cap Gris Nez, even if the service is not to have a more positive role. However the number of collisions is unfortunately only one factor determining the extent of oil pollution. A single casualty involving a big loaded tanker can undo years of careful preventive work. The frightful pile-up off the Varne in 1971 released only a few hundred tons of oil whereas a single collision between the *Olympic Alliance* and HMS *Achilles* in 1975 – otherwise a fairly good year – caused a serious spill.

But whatever qualifications one may add, one-way routeing is obviously a permanent feature of the maritime scene. If one then assumes that at some of the real accident black spots radar surveillance is also going to be cost-effective in reducing casualties, as well as politically desirable, the next question is whether this passive system should become an active one. Could the Coastguard at Langdon Bay ever direct oil tankers through the Dover

Strait as air traffic controllers at West Drayton direct jumbo jets through the London terminal area? And will they actually do so one day?

The answer to the second question is surely No; yet theoretically such a thing is possible and it is worth thinking the possibility through if only to reject it.

The sheer volume of traffic would not be a problem. Heathrow Airport's controllers alone handle nearly twice as much traffic in a day as the Coastguard see on their sceens at Langdon Bay, and it moves a great deal faster. Most of the time, admittedly, the job of separating conflicting traffic in the air is made much easier by the third dimension that is not available at sea – and a disaster like the Zagreb crash in 1976, when a British Airways Trident collided with a Yugoslav DC 9 at 33,000 feet, shows how crucial this height separation is. Yet in the final landing phase airliners do have to be marshalled into an orderly queue – rather like the cross-Channel ferries queueing to enter Calais harbour at the height of the summer season – because they all want to use the same glidepath on to the runway.

If positive shipping control were ever attempted, careful spacing of the traffic would be critical. The aim would be to avoid a close quarters situation the shore controller could not handle. Ships would check in and out of the traffic schemes at pre-arranged times, just as an aircraft files a flight plan before take-off to warn the various control centres of its predicted arrival in their airspace. Ships' masters who turned up early would have to admit their error and turn circles off the Bassurelle light vessel or the Noord Hinder until it was their turn to set off through the Dover Strait, like an airline captain holding in the Bovingdon 'stack' while waiting his turn to land at Heathrow.

The air traffic control system is immensely complex, relying on a constant interchange of telexed messages processed by computers in addition to the direct radio contacts between controller and pilot, radio beacons, coded radar transponders and so on. But it works, and no doubt if the boffins were let loose on the problem of ship control they could design some sort of working system for that too, albeit at great cost and inconvenience to the seafarer. For example all craft entering a controlled shipping lane would have to carry the minimum equipment to use the system – presumably including VHF radio and radar – just as light aircraft must be instrumented almost to an airliner's standard to use London's major airports. Small sailing boats that could not be relied on to respond predictably to the controller's instructions would probably be banished altogether.

It is a disturbing prospect to anyone who values the traditional freedom of the seas. Professional mariners would probably fight to retain their independence of the shore, even though they have generally accepted the principle of traffic routeing as a necessary safety measure. Shipowners would protest at the cost and the extension of administrative bureaucracy. Imagine our sea transport brought to a halt by a shipping controllers' strike! It may seem far fetched, but it happens to air transport.

A reassuring thought is that any scheme of shipping control would have to be negotiated through the slow moving IMCO machinery, and IMCO is notoriously better at comfortable compromise solutions than it is at bold initiatives – especially expensive ones. And then there is the weather.

In this situation gales would be a reactionary old seadog's best friend. A jet airliner handles fairly predictably even when flying through a storm, although it may admittedly have trouble landing in fog. At sea, even a giant supertanker is sometimes at the mercy of the weather, and little coasters still run for shelter from a severe gale along with the passenger ferries. Under these circumstances a rigid system of shore control would be disrupted or paralysed. One flexible enough to cope with the unpredictable would amount to no more than a set of advisory navigational rules which the man on the bridge at sea was left to apply as he judged fit – which is the system we already have.

In short, the positive shore control of Channel shipping analogous to the ground control of aircraft may be possible, but it is almost certainly impracticable as well as undesirable. Langdon Bay and its French counterparts will no doubt continue to develop and in time be accepted as an entirely routine aid to navigation. But only on an advisory or informative basis – which is after all what the names CNIS and SINM imply.

The nagging doubt left by this conclusion is whether the sheer physical capacity of the automatic data processing equipment Decca is installing at Langdon might tempt the Coastguard operators into a more ambitious role. It has been designed to track 250 radar echoes, calculating course, speed and relative positions at the touch of a button. With such a capable system at one's fingertips, why not make more positive use of it?

The formal answer is that the specification was written with nothing more than information in mind. The man who wrote it, the head of CNIS, Captain Richard Emden, would not even accept the adjective advisory. He laid great emphasis on the concept of an 'informative' service, carefully distinguishing that function from the provision of advice and the direct intervention that occurs in the air.

The crucial point here is that the direct control of ships, or even offering a Master navigational advice in a potential collision situation, would carry with them corresponding responsibilities the Coastguard is simply not organised to accept.

Yet even this point does not finally resolve the debate. It is not enough to say that shipping control poses difficult legal and operational problems, or that it would cost a lot of money. Achieving safety in the air is expensive. The present limited service at sea was not created to make jobs for the Coastguard, much as they may appreciate their new role, but in response to the unacceptable risk of continued tanker disasters – the loss of life would not in itself have been enough – and the *Amoco Cadiz* has jerked the public into realising that the risk still exists. If seafarers have to swallow their traditionalist pride and accept some of the aviators' pedantic professional discipline, then so be it.

The big oil companies would probably reply indignantly that in the best run fleets, which include their own, the point has long been taken. A great deal of work has already gone into bridge organisation and navigational planning so as to eliminate the one-man error. Voyage plans are prepared and checked almost like flight plans. Watches are strengthened when manoeuvring through difficult waters like the Dover Strait, engine rooms kept on standby and so on. Casualty reports are already circulated within the tanker industry. In other words, where the lessons of aviation are relevant and practicable, they have been applied.

The flaw in this argument is that even if there were no room for improvement aboard oil company ships, they represent only about forty per cent of the world's tanker tonnage. Most of the ships – including most of the sub-standard ones – are hired either on a time charter, or voyage by voyage on the spot market, to meet fluctuating requirements. That way the oil companies do not carry their own surplus tonnage. They can and do exert some control over the safety of their time chartered ships, studying the owner's record rather as the underwriters do, and probably insisting on certain minimum standards of manning and equipment. But inevitably, in a competitive commercial situation, these are only minimum standards, and when it comes to the last minute voyage charter, even these may have to go by the board. If they did not, the hundreds of sub-standard ships which evidently still exist would find no employment.

In short, the oil companies operate a double standard that can only be justified on economic grounds. Theoretically, there are several ways of improving the situation. They could operate a larger proportion of the tonnage they need. They could refuse to hire any vessel that did not come up to their own high standards – and pass the cost on to the consumer. They could press for faster, tougher action through IMCO to raise standards worldwide, even if a lot of older vessels have to be scrapped.

Unless they do something, one way or the other, they will probably face increasing public demands that the owner of the oil cargo, not the ship-owner, should be held responsible for pollution damage. A clause to this effect was written into Britain's new Merchant Shipping Act, only to be removed as part of a last minute deal to rush the Bill through the House of Lords before Parliament was dissolved in April 1979.

In spite of all the attention the sub-standard tanker is getting, it is not going to disappear from the seas over the next few years. In fact it could well become more of a problem as the cost of operating ships from the traditional maritime countries rises, and with it the pressure to find refuge under a flag where rates are low and corners can be cut without too many questions.

If seafarers want entirely to remove the threat of further interference from the shore, they must demonstrate that they can achieve safer sea transport by other means. One approach that is already being tried is in effect to transfer some of the shore controller's analytical equipment on to the ship's bridge at sea.

A number of manufacturers produce collision avoidance systems for use with radar. They range from what are really no more than automated plotting aids, which record the passage of the radar blip across the screen, to fully automatic, computerised systems which will track echoes and show a 'predicted area of danger' that one's own ship should avoid.

Their use is still intensely controversial. The Americans have typically pushed ahead with a unilateral requirement that tankers entering their waters must carry a collision avoidance system from 1982. The British have shown an equally typical caution, hinting meanwhile that the US initiative may owe more to the strength of the computer industry's lobby in Washington than it does to nautical common sense. The British Government is also concerned – in this case with Soviet support – that whatever requirement is laid down, it should be negotiated through IMCO's international machinery. But IMCO itself has had great difficulty in establishing an agreed technical standard.

Opinion is divided on two main issues: whether the technology is yet good enough to be relied upon in working conditions at sea, and whether complex automatic equipment in some navigators' hands may not create more problems than it solves. Supporters of the American approach argue that even if the equipment is not totally reliable, it must be a safety bonus if it alerts watchkeepers to some dangers of which they were not otherwise aware. The more conservative British analysis suggests that an automatic system improves the performance of a good navigator, but may leave a bad one over-confident or confused, and therefore more dangerous than he was without it.

There are some interesting analogies with aviation here – the cockpit warning systems that get ignored because they so often turn out to be a false alarm, the problem of defining an 'acceptable' failure rate, and so on. But the argument is more directly reminiscent of the early days of radar aboard ships, when its misuse sometimes led to 'radar assisted collisions'.

Some were simply the result of false confidence. Others were caused by misunderstanding the radar display, which in those days only showed other ships' movement relative to one's own position in the middle of the screen. This is admittedly the most important single piece of information a navigator needs to know to avoid a collision, but it is also confusing, because the movement of other ships cannot be read straight off the display. A 'stationary' echo is merely keeping pace with one's own ship.

After an analysis of many previous collisions, Decca therefore produced the first 'true motion' radar display. It was achieved by the brilliantly simple idea of moving the ship's own position across the screen according to the course and speed it was actually making, so that the coastline, if any, stayed put, and other ships' echoes behaved as if the scene was being filmed from a hovering helicopter.

Exactly the same raw data was now being presented but in a different form, which a ship's officer instinctively understood because it was the way he had always mentally plotted his vessel's movement across the chart. The lesson as applied to collision avoidance systems is obvious – that the amount of information the computer supplies is less important than the form in which it is presented.

A more basic approach to tanker safety in which IMCO, Lloyd's, the American Bureau of Shipping, Bureau Veritas and the other ship classification societies have long been involved is to raise the design standards of the ships themselves. To take one of the simplest examples, all tankers ordered since 1971 have been subject to an IMCO limitation on the size of individual cargo tanks, so as to reduce the amount of oil spilled if the ship is holed.

In 1974, the new Safety of Life at Sea (SOLAS) convention revised the whole range of standards, covering construction, stability and installed machinery as well as equipment for life-saving and navigation. And in February 1978, although the new SOLAS convention was not yet in force, the Tanker Safety Conference described earlier revised them yet again for this particular type of vessel. It was decided among other things to make the requirements for duplicated steering gear more explicit. The conference agreed that an alternative power supply at least capable of steering the ship at half speed must be installed in new vessels, and that:

> Each main steering gear shall comprise two or more identical power units and shall be capable of operating the rudders while operating with one or more power units. As far as is reasonable and practicable, the main steering gear shall be so arranged that a single failure in its piping or in one of the power units will not impair the integrity of the remaining part of the steering gear.

It was almost as if delegates had a premonition of what was to happen off Ushant a few weeks later. But as the story of the *Amoco Cadiz* emerged, it became clear that even that careful wording of the requirement was unsatisfactory – at least where complex, cross feeding electro-hydraulic gears were concerned. The Liberian board of investigation's report commented:

> The avoidance of the possibility that one failure may put all the steering arrangements out of action now rests on the rather tenuous phrase 'as far as reasonable and practicable'. It could be held that the steering gear of the *Amoco Cadiz* fulfilled this requirement by virtue of the isolating and crossover arrangements provided. In practice, of course, it did not. Perhaps the complexity of the efforts made to provide numerous alternative modes of operation contributed to the failure in this instance.
>
> Single failures which have nullified the whole steering gear have happened before, and if the avoidance of such loss of manoeuvrability is desired it seems that a return to the simple concept of two completely independent units, each having a separate oil supply and hydraulic system, or of providing a completely independent auxiliary steering gear having a specified performance, would produce a steering gear less vulnerable to complete failure.

Prompted by the French, IMCO has in fact taken another look at it, and at two other aspects of VLCC design which might have some relevance to what happened off their coastline – twin rudders and anchoring equipment.

A vessel with twin screws and rudders should be more manoeuvrable and possibly more reliable than a single screw, single rudder ship. But in the past tanker operators have generally argued that even if both these advantages could be demonstrated in practice – and they question any difference in reliability – the extra cost of duplication would not be justified. Others might think differently. However theoretical studies suggest that in a vessel as bulky as a VLCC, the differential steering effect of twin screws – with no rudders – would be useless in winds of more than fifteen to twenty-five knots, or about Force 5. In other words they would have been no help to Captain Bardari.

On the question of anchoring, the Liberian investigation had a good deal to say, emphasising how little faith the Italian crew had in their ground tackle's ability to save them from the rocks – and how right they were. The report judged the immediate cause of the steam winch breaking was probably using it as a back-up to the brake, which failed to stop cable paying out even when it was tightened up with a hammer, and not taking enough care to drain the steam line feeding it along the deck from the engine room. When the tremendous strain on the port anchor began to override both the brake and the steam pressure in the winch, bolts of water condensed in the deck line must have been forced through the system to hammer the cylinders from their mounting.

The Liberian board commented that a reluctance to rely on the winch brake to hold the anchor cable 'appears to be fairly common on this size of vessel, where the anchors and cable are very large and very heavy and capable of putting a very large strain on the windlass.' Such vessels had to be virtually stopped in the water before any anchor was dropped.

The point about this reluctance, if it existed on the *Amoco Cadiz*, is that it might have discouraged the crew from preparing both anchors for lowering earlier in the day, by removing the wire lashings and stopper blocks. By the time they were needed, the seas breaking over the bows made it too dangerous to work on the starboard side, so only one anchor was usable.

The board felt unable to say whether dropping anchors earlier would have affected the eventual outcome, but implied that it might have helped because the tanker drifted over some comparatively shallow patches. The evidence of the broken anchor itself, on the other hand, with both flukes torn off at the roots and the hole in the stock through which they pivot visibly distorted, tends to bear out Captain Bardari's judgement that dropping it was little more than a gesture.

Yet as an empty gesture we can still learn from it. As the board finally suggested, this casualty may have 'drawn attention to a weak point in the seaworthiness of every VLCC, that the anchoring equipment presently in use is inadequate.' It's sad that just as in the aircraft business so many lessons of this kind are learnt only from the post mortem examination.

Every week, there is usually at least one serious tanker casualty somewhere in the world. And in spite of obvious black periods like 1971 in the Channel, and the American winter of 1976–77, the overall accident rate remains remarkably steady at about two serious casualties for every hundred tankers at risk each year.

SERIOUS CASUALTIES INVOLVING OIL AND CHEMICAL TANKERS
1968–1977

Year	*No. of Casualties*	*Tankers at Risk*	*Rate per 100*
1968	79	2,943	2.68
1969	76	2,991	2.54
1970	60	3,049	1.97
1971	64	3,154	2.03
1972	74	3,217	2.30
1973	65	3,281	1.98
1974	69	3,443	2.00
1975	88	3,553	2.48
1976	93	3,572	2.60
1977	85	3,492	2.43
Totals	753	32,695	2.30

Source: UK Tanker Safety Group

Whether one should despair at finding no real improvement over ten years, or take heart that things have not got worse, is a surprisingly difficult question to answer. As with most statistics, people interpret them to suit their own point of view. For example the international oil companies, concerned for their public image after a big tanker disaster, are fond of breaking these figures down to support their plausible argument that oil transport is 'safer' in giant supertankers than it was in the smaller vessels they replaced. For a start, the number of vessels required is reduced by a factor of at least five. If the United States were to build a few deep-water terminals offshore, their daily crude oil imports could be handled by four VLCCs instead of twenty 50,000 tonners trading into congested harbours like Philadelphia, Houston and New Orleans, or whatever mixture of tonnages is currently being used. With fewer tankers, the argument goes, there should be fewer opportunities for collision, a relatively larger supply of skilled seafarers to man them, and plenty of money available to lavish on their equipment.

At first sight, it is true, the record seems to endorse this logic:

SERIOUS TANKER CASUALTIES BY SIZE OF SHIP 1968–1977

	10,000–25,000 dwt.		25,000–45,000 dwt.		45,000–150,000 dwt.		Above 150,000 dwt.	
Year	*Number*	*Rate per 100*	*Number*	*Rate per 100*	*Number*	*Rate per 100*	*Number*	*Rate per 100*
1968	47	3.38	17	2.26	15	1.91	0	–
1969	40	3.01	15	1.94	17	2.04	4	7.02
1970	33	2.60	8	1.02	16	1.83	3	2.48
1971	20	1.61	21	2.57	15	1.67	8	4.04
1972	32	2.63	18	2.21	17	1.85	7	2.62
1973	24	2.08	19	2.25	20	2.14	2	0.57
1974	23	2.04	19	2.17	22	2.24	5	1.10
1975	27	2.59	23	2.64	28	2.65	10	1.72
1976	23	2.41	29	3.49	30	2.74	11	1.58
1977	14	1.64	28	3.54	33	3.06	10	1.30
Totals	283	2.44	197	2.42	213	2.25	60	1.71

Source: UK Tanker Safety Group

The accident rate for ships of more than 150,000 tons stands out as a lot lower than for the other categories, whereas it could be somewhat higher without necessarily destroying the oil companies' argument. That depends on what they mean by safety.

More than a hundred men a year are killed aboard tankers. Fewer ships and a lower accident rate must reduce that figure. But assessing the comparative pollution risk is much more difficult. A smaller fleet of large ships each carry proportionately more oil, but do they tend to spill proportionately more or less of their cargo when they get into trouble? Would the people of Brittany have preferred four or five wrecked 50,000 tonners (or perhaps six to allow for a higher casualty rate) instead of one *Amoco Cadiz*? It's an impossible question. Which Bretons are we asking anyway – those living in Portsall or in Concarneau?

A more substantial reason for caution about the supertankers' accident record is that they are all relatively new ships and the casualty statistics show that most of the trouble comes from older vessels, just as one would expect:

CASUALTIES BY AGE 1968–1976

Age group in years	*0–4*	*5–9*	*10–14*	*15 and older*
Serious casualty rate	1.6	2.0	1.9	3.1
Total loss rate	0.2	0.2	0.2	0.9

Source: UK Tanker Safety Group

Until we see how the giant tankers fare in their old age, therefore, and who operates them, it seems wiser to leave the question of their comparative safety open.

If all the other factors were equal, the teams who actually have to clean up oil spills would sooner get it over with quickly than face a long slow seepage – which is why they like to blow wrecks up. On the other hand their equipment's capacity is severely limited. The British authorities, for example, can treat about 6,000 tons of oil a day in the Channel. Beyond that, they have to leave it to go ashore. For most of us, no amount of small casualties would have quite the political and emotional impact of a single catastrophe like the loss of the *Torrey Canyon* or the *Amoco Cadiz*.

Another way of breaking the casualty figures down is by flag, so as to produce a sort of safety league table and, commonly, to provide ammunition for attacks on Liberia, Panama and the other countries offering 'flags of convenience':

SERIOUS CASUALTIES BY FLAG 1968–1976

			Rate (percentage)	
Flag	*Number*	*Deadweight*	*Number*	*Deadweight*
France	11	1,276,800	1.3	1.6
Greece	84	3,370,100	3.9	3.5
Italy	12	876,100	1.2	1.8
Japan	17	1,654,500	1.0	0.8
Liberia	209	11,221,400	3.0	2.2
Norway	47	3,016,600	1.9	1.6
Panama	40	1,258,400	2.9	2.0
UK	66	4,135,100	1.9	1.8
USA	53	1,717,200	1.8	2.0
All flags	668	35,802,400	2.2	1.9

Source: UK Tanker Safety Group

The Liberian and Panamanian fleets evidently do have more than their share of casualties, surpassed only by Greece, which is sometimes accusingly described as operating a 'quasi flag of convenience'.

The conveniencc flags are those that do not have a long, indigenous maritime tradition behind them. They have been offered, mainly since the

Second World War, literally as a convenience to foreign shipowners who want to avoid the taxation, manning restrictions and other penalties imposed on vessels registered in their own ports. But they are by no means peripheral to international shipping operations. They account for nearly a third of the world's tonnage, and for some years now the Liberian fleet has been the largest in the world, consisting mainly of tankers and bulk carriers. To read the names of the home ports painted on the sterns of tankers pounding up and down the Channel one would think that Monrovia was one of the great maritime trading centres instead of an obscure tropical harbour on the Grain Coast of West Africa.

The safety standards set by governments operating such flags are therefore of crucial importance, and they are not going to develop as a result of the natural process of seafaring experience, as they have done in Britain or the United States. They have to be adopted as a deliberate act of managerial policy. Hence the value, mentioned earlier, of involving such governments as closely as possible with IMCO.

The significant convenience flags these days are Liberia, Panama, Cyprus and Singapore. The shipowners who use them include a lot of Greeks – although the Greek Government has recently been trying to attract tonnage back to its own flag – plus Americans, Japanese, West Germans, Scandinavians and a few British. Major oil companies, which operate their own fleets as well as chartering from independent owners, are among the customers.

In the years after *Torrey Canyon*, the Liberian flag particularly came in for widespread criticism, much of it undoubtedly justified by the facts, some of it mere prejudice in the search for a scapegoat. With so many ships flying the West African Stars and Stripes they were bound to figure prominently in the casualty lists. Yet it is a matter of history that some of the most notorious names in terms of oil pollution have that flag in common.

At any rate the criticism eventually got home, both to the Liberian administration and to reputable shipowners who found it expedient to register their ships abroad but had to justify their actions elsewhere. In 1971 the Liberian Bureau of Maritime Affairs set up its own world-wide inspection service, with 150 agents, to see that its fleet was maintained and equipped to international standards.

In other words it was trying to meet the most fundamental criticism of convenience flag governments – that they take no direct responsibility for the ships whose registration fees they accept. Whether the inspection service actually achieved that objective is another matter. When the *Argo Merchant* was lost in 1976, and found to be six months overdue for her annual inspection, it was alleged that thirty per cent of Liberian vessels were being neglected in this way. That winter's shocking casualties produced a new initiative from the Liberian Shipping Council, representing owners who use the flag, and the Maritime Bureau responded with a tougher programme of inspection without notice and detention of ships not meeting the necessary standard. Vessels more than twenty years old were to be checked at least every nine months and those with a record of deficiencies every six months (since 1975 ships over twenty years old have not been accepted for registration).

By the time the *Amoco Cadiz* went ashore, the number of inspection agents had been increased to 200, and the Maritime Bureau maintained that in many respects Liberian safety requirements were higher than the IMCO standard. But the flag's dubious image had evidently not been eradicated. The interim report of the subsequent public inquiry opened with the claim that it was conducted 'in an atmosphere of prejudice' encouraged by hostile press and television references to flags of convenience. According to the board of investigation:

> The facts are very different from what the public was led to suppose. The *Amoco Cadiz* was in fact a relatively new vessel, built only in 1974, in a Spanish shipyard of high reputation, and she was equipped with all the most modern aids to navigation, being classed in the highest class of one of the most reputable classification societies. . . . It soon became clear to the board that the causes of this lamentable casualty were not to be found in the fact that the *Amoco Cadiz* happened to wear the Liberian flag, or that she was an ill-equipped or ill-manned vessel.

As far as this ship was concerned, the statement is irrefutable, but the broader argument will only be settled by the long term trend of the casualty statistics. Meanwhile, to be fair to the critics, the attacks on flags of convenience have themselves been broadened recently to take in all 'sub-standard' ships, whatever flag they fly. One reason is to include in their strictures the Greeks, some of whom have a reputation for running ships on the cheap, and whose own flag has in the past had a worse record of tanker safety than either the Liberian or Panamanian flags.

An important study of sub-standard tankers by F. M. van Poelgeest, publishd by the Netherlands Maritime Institute in February 1978, went so far as to identify about 450 vessels in this category, of which approximately 175 were sailing under the Greek flag and a further substantial number were managed by Greeks though flying the Liberian flag. His central conclusions were:

(1) that the proportion of accidents under the Greek, Liberian, Panamanian and Cypriot flags was above average;
(2) that tankers owned by oil companies were much less likely to give trouble than those independently owned;
(3) that about half the casualties were concentrated among small, elderly vessels (between about 15,000 and 30,000 tons and over sixteen years old).

The main cause of the higher casualty rates was identified as poor manning; and 'human operating errors' were claimed to be responsible for three quarters of the incidents.

This last analysis is a difficult one. Similar statements are often made about aircraft accidents but pilots complain bitterly that they are made the scapegoats for failures elsewhere in the industry, simply because an error by a pilot – usually a dead pilot afterwards – closed a loop initiated by bad design, maintenance or management.

Yet even with that proviso, the human element is clearly of great importance at sea, especially in the collisions and groundings that cause so much of the accidental oil pollution. This was belatedly recognised by IMCO in July 1978, when the United Nations agency produced its first convention on training, certification and watchkeeping standards. No one pretended that the new convention would do more than set the absolute minimum qualifications for masters, mates, engineers and radio operators, a lowest common denominator that any responsible administration should demand; and the usual delay before it came into force was expected. But it was nonetheless a valuable step forward.

Unfortunately, there was a sad little backwash from all the diplomatic work that went into its preparation, in that the British Goverment felt obliged to close the legislative loopholes in its own system of certification before demanding that others did the same. In principle this was to be commended, of course, but in practice it meant that a coastal seaman who had half a lifetime's hard experience navigating, say, a sand barge through the channels of the Thames Estuary, was now expected to produce paper qualifications. How many competent seafarers will be lost to coastal shipping as a result remains to be seen. There are temporary loopholes. But the fact is that yet another small area of sea-going freedom has been curtailed.

Shortly before van Poelgeest published his study for the Netherlands Maritime Institute, representatives of eight European governments (Belgium, Denmark, Federal Germany, France, the Netherlands, Norway, Sweden and the UK) met in the Hague to sign a memorandum on joint action against sub-standard ships in the North Sea. Another crucially important principle was involved here – that coastal states should take more responsibility for protecting their own shores from accidents and pollution by rigorously inspecting ships in port, whatever their nationality. Many authorities, including the Americans, whose Coastguard is quite prepared to take a unilateral lead in safety enforcement, argue that more will be achieved in this way than by relying on flag states which have little direct incentive to regulate their own shipping.

France, too, has shown since March 1978 that she is willing to take the international law into her own hands. But willingness is not enough. Some governments simply do not have a body of surveyors to undertake this inspection work other than by hiring the services of a classification society, and one that does, like Britain, may well have difficulty recruiting the skilled men it needs in competition with the shipping industry.

This same problem of allocating scarce resources crops up when people call for compulsory pilotage in dangerous waters like the Channel. It's obviously an attractive idea. Indeed Britain's new Merchant Shipping Act of 1979 made provision for the extension of compulsory pilotage where local conditions warranted it. But to apply it throughout British waters would mean diverting thousands of the most highly skilled officers from running their own ships.

In both these cases it's a question of the price we are prepared to pay for safety. Judging by the vast sums people claim for pollution damage one would think they were ready to pay more than they have done. How much more no doubt depends on when the next *Amoco Cadiz* hits the beach. Meanwhile the practice of port state inspections – which is after all embodied in IMCO's 1960 SOLAS convention – is likely to gain ground. The Greek Goverment has already said it wants to adopt the Hague memorandum.

However the Greek flag was back in an unpleasant limelight in July 1978 when London's cargo underwriters increased the surcharge for ships over fifteen years of age flying certain flags. Greek owners were aghast to find their own flag bracketed with Cyprus, the Dominican Republic, Liberia and Somalia on the underwriters' black list. They protested that there was a danger of 'burning the green grass with the dry grass'. But most outsiders saw the move in a different metaphor, as an encouraging sign that the commercial worm was at last turning. If sanctions could be applied in this direct fashion within the shipping industry, a lot of expensive international bureaucracy might be avoided.

Where hull and machinery insurance is concerned, a basic sanction has long been applied through the ship classification societies, in that cover is difficult to obtain for a ship not 'classed' as in sound condition – 100A1 at Lloyd's, or the equivalent standard with the American Bureau of Shipping, Bureau Veritas and so on. The underwriting syndicates also vary the rates they charge by a factor of perhaps five according to the vessel's age and the owner's record (a confidential index, or 'green book', is maintained at Lloyd's for this purpose). However underwriters are the first to acknowledge the limits of their own discrimination in an intensely competitive international market. It's a case of 'If you don't want the business I know somebody else who will.' The net effect is probably that operators of modern, well-maintained tonnage are subsidising the sub-standard fleets.

When everything else fails, owners and insurers both have to put their trust in the Captain Weinerts of this world, lurking near the maritime danger spots in the hope of picking up a fat salvage award.

Like compulsory pilotage, the idea of governments providing a safety net of publicly funded salvage tugs has an obvious appeal. Part of the French Government's response to *Amoco Cadiz* was to sign the contract to keep the *Abeille Normandie* on station at Brest all the year round. The British firm United Towing tried to interest its own government in a similar scheme, but Whitehall took the view that the waters of North West Europe were already relatively well supplied with large tugs and settled instead for the purchase of two caches of portable salvage equipment like the submersible pumps that proved so useful in lightening the *Christos Bitas*.

More salvage tugs capable of manoeuvring a VLCC would obviously help, but no one pretends that they provide an answer to tanker casualties. The next disaster may not occur either in the Channel or the Irish Sea, but perhaps off the Shetland Isles, at the new Sullom Voe oil terminal, where there have already been angry complaints about pollution. If the salvage skippers knew where it would strike next, they would already be there.

One longer term move to improve the prospects of salvage did get under way in London, however. Tanker owners, underwriters and salvage companies sat down to devise a new contract to replace the traditional Lloyd's Open Form – or Lloyd's Standard Form as it is nowadays officially known – over which Weinert and Bardari wasted so many angry words.

The formula has served seafarers well for many years all over the world. But all sides of the industry were agreed that it could probably be improved to take more positive account of a salvage operation's value in preventing pollution as well as saving ship and cargo – indeed of the need sometimes to put the prevention of an oil spillage above every other consideration. There was also a growing need for salvage tugs to operate under government direction rather than on a commercial 'no cure – no pay' basis, because coastal states threatened by pollution were increasingly intent on intervening to prevent that type of casualty even at the expense of the traditional right of 'innocent passage' through another country's territorial waters.

Apart from Abeille, Bugsier and United Towing, the big European names in salvage are the Dutch firms of Smit and Wijsmuller. The North Sea is therefore well covered from the base ports of four of these companies – Hamburg or Cuxhaven (Bugsier), Ijmuiden (Wijsmuller), Rotterdam (Smit) and Hull (United Towing). The French concern covers the southern side of the Channel, sometimes with a Bugsier tug for company, and when the Dutch boats are not busy towing oil rigs they can often be found waiting at Dover or Land's End on the English shore, with an interested eye on the weather and a twenty-four-hour radio watch for that first crackling distress call.

The deep sea salvage business still rewards enterprise in its purest form. It is unjust and misleading to describe the big tugs as 'the vultures of the sea' – which some were inclined to when they heard the story of the *Amoco Cadiz* – because if a ship and her crew are in real trouble nothing is more welcome. Yet they are something of a romantic anachronism, a throwback to an older, more elemental tradition of seafaring where greed and altruism are mingled in the stark gambler's formula of 'no cure – no pay'. They represent the sort of freedom that is everywhere being lost at sea.

Today, if you stand on the high ground beside the Portsall lifeboat house, as Jean Gouzien did on the evening of 16 March 1978, there is nothing but empty sea where the *Amoco Cadiz* went down. But the effects of the disaster will be felt for many years yet.

On the positive side, many lessons have clearly been learnt – most of them at the Breton people's expense. The slow moves towards better designed, better equipped and better manned tankers which began in the wake of the *Torrey Canyon* have been accelerated by this new casualty. A few basic principles for coping with oil spills have been established.

But it is equally clear that the preventive measures so far taken will not be enough to stop the same thing happening again, this time perhaps deep inside the Channel's bottleneck, where the tides swirl past the shifting Goodwin Sands, in the approaches to the North Sea oilfields' terminal at Sullom Voe or in one of a dozen places along the eastern seaboard of the Unites States. Not that the tanker business can ever be entirely free from accident. The sea is a dangerous place. But no particular tanker accident is inevitable.

Over the past few years a great deal of fuss has been made about steeply rising oil prices and their effect on the industrialised economies of the West.

But there is an added price for fuelling those economies which is never entered in the oil company accounts because it cannot readily be measured in pounds, francs or dollars. How do you cost the miserable death of one oiled sea bird? Or of a thousand sea birds? Does it matter to most of us that the traditional Breton oyster is replaced on the menu by some strange new variety, or that if you dig beneath the bright yellow sand you find a dark layer of residual oil?

Yet until such things are costed the accountants who organise our greedy economic lives will take no notice. And we in turn shall not be prepared to pay the full price of preventing those tanker casualties that can reasonably be prevented, and providing proper means to clean up after them.

As things stand, it is only a matter of time before the black tide rises again.

Index

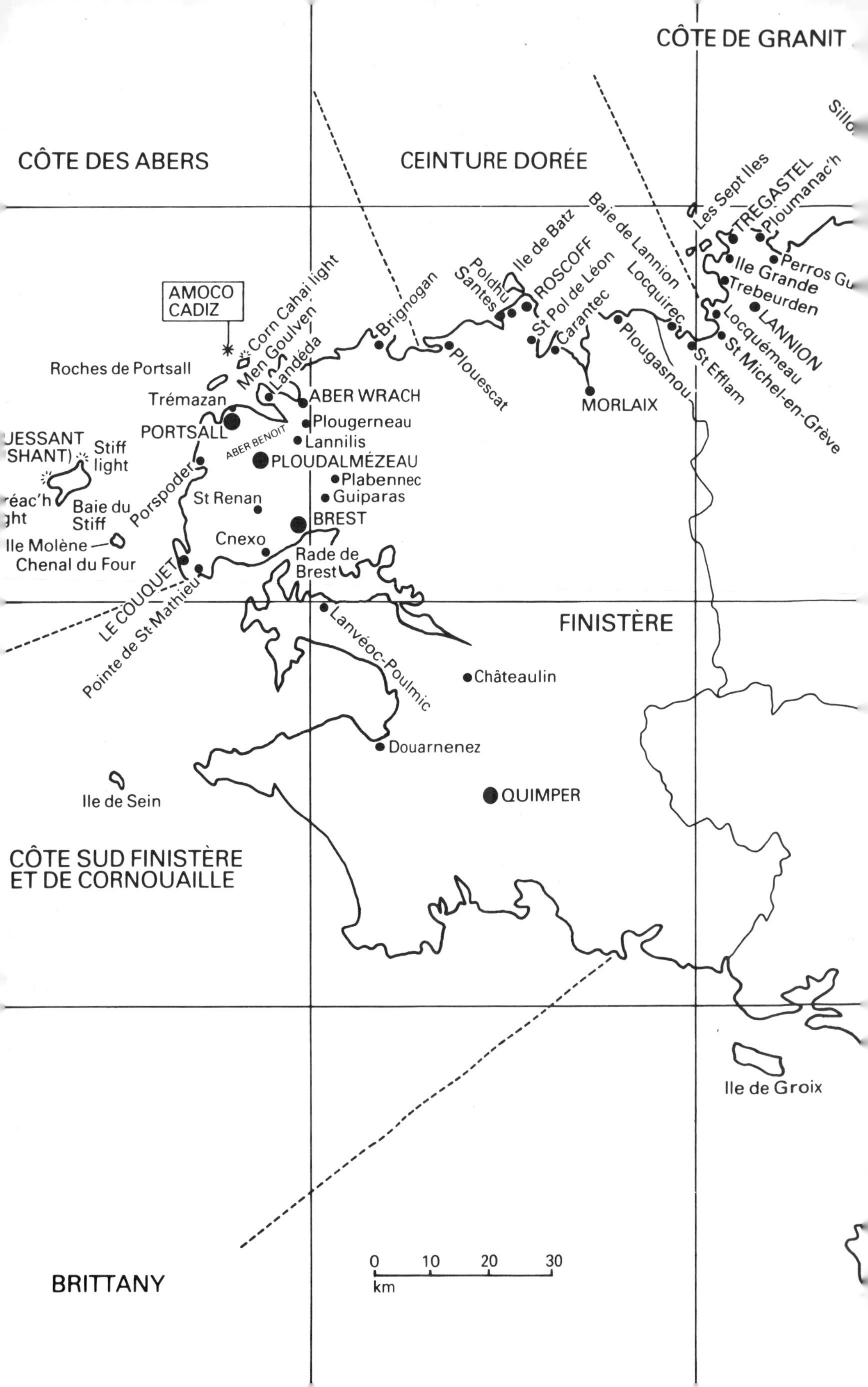
CÔTE DE GRANIT
CÔTE DES ABERS
CEINTURE DORÉE
AMOCO CADIZ
Corn Cahai light
Men Goulven
Landéda
Roches de Portsall
Trémazan
PORTSALL
ABER BENOIT
ABER WRACH
Plougerneau
Lannilis
PLOUDALMÉZEAU
Plabennec
Guiparas
Stiff light
Baie du Stiff
Porspoder
St Renan
BREST
Ile Molène
Chenal du Four
Cnexo
Rade de Brest
LE COUQUET
Pointe de St Mathieu
Lanvéoc-Poulmic
Brignogan
Plouescat
Poldhu
Santes
Ile de Batz
ROSCOFF
St Pol de Léon
Carantec
MORLAIX
Plougasnou
Baie de Lannion
Locquirec
St Efflam
Les Sept Iles
TREGASTEL
Ploumanac'h
Perros Gu
Ile Grande
Trebeurden
LANNION
Locquémeau
St Michel-en-Grève
FINISTÈRE
Châteaulin
Douarnenez
QUIMPER
Ile de Sein
CÔTE SUD FINISTÈRE ET DE CORNOUAILLE
Ile de Groix
0 10 20 30
km
BRITTANY